Marine Emergenc

This book is an influential guide to marine emergencies and the current strategies that can be employed to cope with the immediate after-effects and ramifications of disaster at sea. Many mariners will at some point in their marine careers become involved in one sort of emergency or another, while in port or at sea, whether it is a fire on board, a collision with another vessel or an engine failure threatening a lee shore. Actions to take in such incidents can be the difference between survival and catastrophic loss.

This text provides a direct insight into some of the latest incidents and includes:

- case studies from emergencies worldwide
- checklists and suggestions for emergency situations
- everything from fire and collision right through to the legal implications of salvage.

D.J. House has written and published 18 marine titles, many of which are in multiple editions. After commencing his seagoing career in 1962, he was initially engaged on general cargo vessels. He later experienced worldwide trade with passenger, container, ro-ro, reefer ships and bulk cargoes. He left the sea in 1978 with a Master Mariner's qualification and commenced teaching at the Fleetwood Nautical College. Retiring in 2012 after 33 years of teaching in nautical education, David House continues to research and write for the ever-changing marine industry.

Other Works Published by D.J. House

Seamanship Techniques, combined volume (4th edition), 2013, Routledge.
ISBN 9780415829526 (hbk), 9780415810050 (pbk), 9780203796702 (ebk)

Seamanship Techniques Volume III: 'The Command Companion', 2000, Butterworth/Heinemann.
ISBN 0750644435

Marine Survival (3rd edition), 2011, Witherby Publishing Group.
ISBN 9781856093552

Navigation for Masters (4th edition), 2007, Witherby Publishing Group.
ISBN 1856092712

An Introduction to Helicopter Operations at Sea: A Guide for Industry (2nd edition), 1998, Witherby.
ISBN 1856091686

Cargo Work (7th edition revised), 1998, Butterworth/Heinemann.
ISBN 0750665556

Anchor Practice: A Guide for Industry, 2001, Witherby Publishing Group.
ISBN 1856092127

Marine Ferry Transports: An Operators Guide, 2002, Witherby Publishing Group.
ISBN 1856092313

Dry Docking and Shipboard Maintenance, 2003, Witherby Publishing Group.
ISBN 1856092453

Heavy Lift and Rigging, 2005, Brown Son & Ferguson.
ISBN 0851747205

The Seamanship Examiner, 2005, Elsevier.
ISBN 075066701X

Ship Handling, 2007, Elsevier.
ISBN 9780750685306

The Ice Navigation Manual, 2010, Witherby Publishing Group.
ISBN 9789053315989

Elements of Modern Ship Construction, 2010, Brown Son & Ferguson.
ISBN 9780851748146

Also:

Marine Technology Reference Book (Safety Chapter), 1990, edited by Nina Morgan, Butterworths.
ISBN 0408027843

Marine Emergencies

For Masters and Mates

D.J. House

Routledge
Taylor & Francis Group

LONDON AND NEW YORK

First published 2014
by Routledge
2 Park Square, Milton Park, Abingdon, Oxon, OX14 4RN

and by Routledge
711 Third Avenue, New York, NY 10017

Routledge is an imprint of the Taylor & Francis Group, an informa business

British Library Cataloguing in Publication Data
A catalogue record for this book is available from the British Library

Library of Congress Cataloging-in-Publication Data
House, D. J.
Marine emergencies / D.J. House.
pages cm
Includes bibliographical references and index.
1. Ships--Safety measures. 2. Merchant marine--Safety measures. 3. Marine
accidents. 4. Ships--Fires and fire prevention. 5. Seamanship. I. Title.
VK200.H66 2014
363.12'3--dc23
2013048962

ISBN13: 978-1-138-02045-0 (pbk)
ISBN13: 978-1-315-77069-7 (ebk)

Typeset in Sabon
by Fakenham Prepress Solutions, Fakenham, Norfolk NR21 8NN

Contents

Acknowledgements   ix

About the Author   xi

Abbreviations   xiii

Terminology and Definitions Associated with Marine Emergencies   xix

1   Collision (Ship to Ship) at Sea   1
    Introduction   1
    Collision: Immediate Effects   1
    Impact Damage   4
    Legal Actions of the Master (Under the Merchant Shipping Act)   5
    General Actions Following Initial Response to a Collision   6
    Incorporation and Use of Checklists   7
    Communications Following a Collision   9
    Voyage Data Recorder (Black Box Recorder)   9
    Note of Protest   11
    The Role of the Ship's Chief Officer (in the Aftermath of Collision)   12
    Collision Patch Construction   13
    Collision Patch Materials   13
    Port of Refuge   14
    Port of Refuge and General Average   15
    Damage Control Parties   16
    Passenger Ship Collision   17
    Tanker Collision   19
    Collision: Typical Damage/Repair Assessment (Hypothetical)   20
    Summary   22

2   Taking the Ground: Grounding, Beaching, Stranding and Docking   25
    Introduction   25
    Running Aground   26
    Incident Report: Loss of the *Riverdance*   28
    Incident Report: *MSC Napoli* (Container Vessel)   31
    Beaching   32
    Grounding/Beaching Summary   34

Case Study: Running Aground   35
Immediate Actions   36
Soundings and Use of Lead Line   37
Emergency Dry Docking   39

3 **The Lee Shore and the Use of Emergency Anchors**   49
Introduction   49
What is the Lee Shore?   50
Loss of Steering   50
Steering Gear Operations   51
Lee Shore: Loss of Main Engine Power   57
Master's Options   57
Relevant Anchor Work   59
Example Stern/Kedge Anchor   70
Kedge Anchor   71
Chain Cable/Stud Link: General Information   72
Heavy Weather Encounter   73
Case Study: The Loss of the *M.V. Braer* (89,730 dwt)   78

4 **Fire on Board**   81
Introduction   81
The Outbreak of Fire on Board the Ship   82
Fire Support Units   84
Fire Parties   84
Firefighting Teams   85
CO_2 Maintenance   86
Security against Accidental Release of CO_2 87
Example Fires   87
Case Incident   92

5 **Abandonment**   105
Introduction   105
Loss of the Ship   106
The Aftermath of the *Herald of Free Enterprise*   107
The Loss of *Costa Concordia*, Passenger Cruise Ship   107
Abandonment Psychology   113
Passenger Behaviour   117
Incident Report   118
Exposure to Risk   118
Evacuation by Free Fall Lifeboat   120
Totally Enclosed Lifeboats   121

Maintenance Programme for Life Saving Appliances | 121
Evacuation by Inflatable Liferaft | 121
Evacuation by Davit-Launched Liferaft | 123
The Role of Rescue Boats in Abandonment | 124
Rescue Boat Operations | 125
Example Liferaft Operations | 126
Evacuation by Means of Marine Evacuation Systems | 126
Shipboard Emergency Drills | 127
Evacuation by Helicopter | 128
Helicopter Operations | 129
Helicopter/Shipboard Operations | 131
Surface-to-Air Medical Evacuation (MediVac) | 132
Helicopter Hi-Line Capabilities | 133
Helicopter Incident Report | 134
Miscellaneous Facts (Related to an Abandonment Situation) | 134

6 **Marine Pollution** | **137**
Introduction | 137
Terminology and Definitions affecting Tanker and Gas Carrier Vessels | 138
Pollutants Other than Oils | 145
The Causes of Maritime Pollution | 146
The Design of the Oil Tanker | 148
Oil Tankers | 149
Pipeline Connections | 150
Anti-Pollution Measures | 150
Oil Spills | 151
Exxon Valdez, 23 March 1989 | 152
Lightening Operations (Ship-to-Ship Transfer) | 153
Ship-to-Ship Oil Transfer | 154
Recovery of Floating Oil Pipelines | 155
Oil Movement | 157
Incident Report: Grounding of the Drilling Rig *Kulluk*, 30 December 2012 | 158
Oil Recovery Equipment | 158
Ballast Water Movement | 159

7 **Towing and Salvage Hazards** | **161**
Introduction | 161
Tug Operations | 162
Harbour and Port Authority Tugs | 163
Oceangoing Salvage Tugs | 163
The Work of the Towmaster | 164

Tugs and Emergency Towing   165
Tug Approval Surveys   166
Cargo Deck Barges (Pontoons)   169
The Insurable Risk   169
Sheer Legs in Salvage Use   172
Salvage Contact   173
Quality of Information   176

8 Miscellaneous and Routine Leading to Potential Hazards   177
Introduction   177
Enclosed Space Entry   177
Fog Encounter   178
Dangers Associated with Restricted Visibility   179
Doubling Watches   180
Ice Navigation   181
Man Overboard (MoB)   183
Example Turning Manoeuvres   184
Rescue Boat Activity   186
Boarding or Disembarking Marine Pilots   186
Navigational Pitfalls of ECDIS   187
Search Patterns Associated with IAMSAR   188
Determination of Track Space   190
Duties of the On-Scene Coordinator   191
Example Checklists   192
The Activities of the US Coast Guard   193
Emergency Communications   194
The Use of Distress Signals   196

Annex 1: Question and Suggested Answers for Senior Officers: Towards Marine
   Examinations   199

Annex 2: Notable Shipping Incidents   211

Annex 3: Lloyd's Standard Form of Salvage Agreement: No Cure - No Pay   213

Annex 4: Lloyd's Standard Form of Salvage Agreement: Salvage and Arbitration Clauses   219

Annex 5: Lloyd's Standard Form of Salvage Agreement: Procedural Rules   225

Annex 6: International Salvage Union: Sub-contract (Award Sharing) 2001   227

Summary   235

Bibliography   237

Index   239

Acknowledgements

Brown Son & Ferguson, Ltd, Marine Publications
Bruce Anchors Ltd
Dubai Dry docks
Fleetwood Nautical Campus of the Blackpool & Fylde College
I. C. Brindle
International Salvage Union
US Coastguard
Viking A/S Nordisk Gummibadsfabrik
Lloyd's List
Smit Maritime Contractors, Europe and Smit International
LOF and its supporting documents have been reproduced with the kind permission of Lloyd's

Additional Photography

Mr G. Edwards Ch/Eng., retd, MN.
Mr J. Bateman, Chief Officer, MN.
Mr S. Mooney, Chief Officer, MN.
Mr J. Leyland, Nautical Studies Lecturer.
Mr M. Ashcroft, Nautical Studies Lecturer.
Mr S. Bateman, Chief Officer, MN.
Mr Peter M. Stacey, Marine Pilot
Mr G. Swindlehurst, Chief Officer, MN.
Mr D. MacNamee, Master Mariner, MN, FNI.

IT Consultant

Mr. C. D. House

About the Author

David House is a Master Mariner, starting his marine career in 1962, until the present day. He spent 15 years at sea on various ship types, from passenger liners to dredging operations, engaged in worldwide trades. His marine experience, including a limited time on warships, was gained aboard general cargo vessels, container ships, roll-on–roll-off (ro-ro) ferries and passenger liners. During his working life at sea he carried a variety of cargoes, including both dry and liquid products, reefer commodities, heavy lifts, containers, vehicles, bulk commodities and timber products.

His time in a seagoing capacity involved him in several real-time emergency situations which have been reflected within this work.* Engaged on worldwide trade he encountered considerable ice experience both in the Baltic and on the North Atlantic winter trades. Heavy weather, fog, tropical storms and a lack of under-keel clearance became influencing factors in his continued writings for the marine industry.

Figure 0.1 The author Mr D. J. House.

His later years were engaged in lecturing to marine students on most maritime disciplines. During this period of over 30 years, he successfully wrote 17 textbooks covering such topics as dry docking, anchor work practice, ferry transports, general seamanship, navigation, ship construction, heavy lift and rigging, cargo work, ship handling, marine survival and helicopter operations.

He continues to work within the marine education arena, teaching and carrying out ongoing research into a variety of marine-related topics. This current work has been enhanced by continued work with the International Institute of Nautical Surveyors, the Fleetwood Nautical Campus of the Blackpool & Fylde College and colleagues within the maritime industries.

NB. Merchant Navy officers are expected to wear many hats on different occasions, sometimes being a navigator, medical officer, cargo officer or naval architect. On some unusual occasions even being a cook to a legal counsellor.

Note

* During his seagoing career, the author had firsthand experience of being aboard a ship running aground in fog and ice conditions on a voyage towards Montreal. His vessel was also torn away from the quayside by fast-flowing ice drifts. His ship was later to part all its mooring ropes and was cast adrift without power, in the restricted waters of the St Lawrence River, Canada.

He later experienced a head-on collision off the Northern Ireland coastline while aboard a ro-ro vessel. This particular incident caused contact with a cliff face at 16 knots. The subsequently damaged vessel was then taken with tug assistance to Belfast dry dock for major repairs.

His experiences also include a fishing boat rescue in the Irish Sea and two fires on board different ships. As the acting medical officer at the time he dealt with mental health problems in crew members and violent outbreaks among personnel, resulting in disciplinary procedures having to be taken.

Abbreviations

A.C.	alternating current
ABS	American Bureau of Shipping
ACAS	Advisory, Conciliation and Arbitration Service
ACGIH	American Conference of Government Industrial Hygienists
AHV	anchor handling vessel
AIS	automated identification system
AMVER	Automated Mutual-Assistance Vessel Rescue System
API	American Petroleum Institute
APM	anchor position mooring
APP	aft perpendicular
B & V	Blohn +Voss Industrietechnik GmbH
B	representative of the ship's centre of buoyancy
B/A	breathing apparatus
BHP	brake horse power
BIMCO	Baltic and International Maritime Council
BL	breaking load
BP	bollard pull
BS	breaking strength
BT	ballast tank
BV	Bureau Veritas
C of A	certificate of approval
C of G	centre of gravity
CBT	clean ballast tank
CD	chart datum
CG	Coast Guard
CMI	Comité Maritime International (International Maritime Committee)
CO_2	carbon dioxide
CSM	cargo securing manual
CSP	commencement of search pattern
CSS	Cargo Stowage and Securing (code)
CSWP	Code of Safe Working Practice
D.C.	direct current
DNV	Det Norske Veritas
DP	dynamic positioning
DPA	designated person ashore

DSC	digital selective calling
DSV	diving support vessel
DWA	dock water allowance
dwt	deadweight tonnage
ECDIS	Electronic Chart and Display Information System
EEBDs	emergency escape breathing devices
EFSWR	extra flexible steel wire rope
EPIRB	emergency position indicating radio beacon
ETA (i)	estimated time of arrival
ETA (ii)	European Tugowners' Association
ETA (iii)	emergency towing arrangement
ETV	emergency towing vessel
EU	European Union
F.O.	fuel oil
FLOFLO	float on, float off
FPSO	floating production, storage and offload vessel
FRC	fast rescue craft
FSE	free surface effect
FSWR	flexible steel wire rope
FW	fresh water
G	representative of a ship's centre of gravity
GA	general average
GHz	gigahertz
GL	Germanischer Lloyd
GM	metacentric height
GMDSS	global maritime distress and safety system
GMT	Greenwich Mean Time
GPS	global positioning system
grt	gross registered tonnage
GZ	righting arm (righting lever in stability)
HDFD	heavy duty floating derrick
HF	high frequency
HLO	helicopter landing officer
HMCG	Her Majesties Coast Guard
HMPE	high molecular weight polyethylene
HNS	hazardous and noxious substances
HP	high pressure
HSE	Health & Safety Executive
HSSC	Harmonised System of Survey and Certification
HW	high water

IACS	International Association of Classification Societies
IALA	International Association of Lighthouse Authorities
IAMSAR	International and Aeronautical Maritime Search and Rescue
IAPPC	International Air Pollution Prevention Certificate
IGS	inert gas system
IMDG	International Maritime Dangerous Goods (code)
IMO	International Marine Organisation
INS	integrated navigation system
IOPP	International Oil Pollution Prevention (MARPOL Certificate)
IPS	integrated power system
ISGOTT	International Oil Tanker and Terminal Safety Guide
ISM	International Safety Management (code)
ISO	International Organisation of Standardisation
ISPPC	International Sewage Pollution Prevention Certificate
ISU	International Salvage Union
IUA	International Underwriting Association
IUMI	International Union of Marine Insurers
IWS	in water survey
K	representative of the position of a ship's keel
kg	kilograms
kHz	kilohertz
kNs	kilonewtons
kts	knots
kW	kilowatt
LAT	lowest astronomical tide
LBP	length between perpendiculars
LFL	lower flammable limit
LOA	length overall
LOF	Lloyd's Open Form of Salvage
lo-lo	load on–load off
LP	low pressure
LPG	liquid propane gas
LR	Lloyd's Register
LRS	Lloyd's Register of Shipping
LSA	life saving appliances
LSSA	Lloyd's Standard Salvage and Arbitration
LW	low water
M (i)	metres
M (ii)	metacentre
	representative of the metacentre

M.V.	motor vessel
MA	mechanical advantage
MAIB	Marine Accident Investigation Branch
MARPOL	International Convention for the Prevention of Oil Pollution
MBL	minimum breaking load
MCA	Maritime Coastguard Agency
MCTC	moment to change trim by 1 cm
Medivac	medical evacuation
MEPC	Marine Environmental Protection Committee
MES	Marine Evacuation System
MF	medium frequency
MFAG	Medical First Aid Guide
MGN	marine guidance notice
MHWN	mean high water neaps
MHWS	mean high water springs
MHz	megahertz
MIN	marine information notice
MLWN	mean low water neaps
MLWS	mean low water springs
mm	millimetres
m/m	mass by mass
MoB	man overboard
MoD	Ministry of Defence
MODU (MOU)	Mobile Offshore Unit
MPCU	Marine Pollution Control Unit
MRCC	Marine Rescue Co-ordination Centre
M/S	Merchant Shipping Act
MSC (i)	Marine Safety Committee (of IMO)
MSC (ii)	Mediterranean Shipping Company
MSI	maritime safety information
MSL	maximum securing load
MSN	merchant shipping notice
MSR	mean spring range
MTSA	Marine Transport Security Act (US)
MW	megawatt
NP	national publication
NBDP	narrow band direct printing (telex)
NFU	non follow up
NLS	noxious liquid substances
NRV	non-return valve

NUC	not under command
OBO	oil, bulk, ore carrier
OCIMF	Oil Companies International Marine Forum
OLB	official log book
OOW	Officer of the Watch
OPIC	Oil Pollution Insurance Certificate
ORB	oil record book
OSC	On Scene Coordinator (Military On Scene Commander)
OSHA	Occupational Safety and Health Administration
P & I	Protection and Indemnity Association
P/V	pressure vacuum
PEL	permissible exposure limit
PHA	preliminary hazard analysis
PIC	person in charge
PNG	pressurised natural gas
PPM (ppm)	parts per million
PRS	Polish Register of Shipping
PSC	Port State Control
psi	pounds per square inch
RD	relative density
RINA	Registro Italiano Navale (Classification Society – Italy)
RNLI	Royal National Lifeboat Institution
ro-pax	roll-on–roll-off passenger vessel
ro-ro	roll-on–roll-off
ROV	remote-operated vehicle
s.h.p.	shaft horse power
SA	Salvage Association
SAR	search and rescue
SART	search and rescue radar transponder
SBE	stand-by engines
SBM	single buoy mooring
SCOPIC	Special Compensation Protection and Indemnity Clause
SCR (i)	Shipowners Casualty Representatives
SCR (ii)	Special Casualty Representative
SF	stowage factor
SL	summer load line
SLS	serviceability limit state (design condition)
SMC	Safety Management Certificate
SMS	safety management system
SOLAS	Safety of Life at Sea (convention)

SOPEP	Ship's Oil Pollution Emergency Plan
SSA	Ship Building and Repair Association
SSHP	Site Safety and Health Plan
STEL	short-term exposure limit
SU	search unit
SW	salt water
SWL	safe working load
SWR	steel wire rope
T (t)	tonnes
Te	tug efficiency
TEU	twenty-foot equivalent unit (container)
TF	tropical fresh
TLV	threshold limit value
TPC	tons per centimetre
TPR	towline pull required
TVAS	Towing Vessel Approvability Scheme
TWA	time-weighted average
UFL	upper flammable limit
UHMPE	ultra-high molecular mass polyethylene
UK	United Kingdom
UKC	under-keel clearance
UKSTC	United Kingdom standard towing conditions
ULC	ultimate load capacity
ULCC	ultra-large crude (oil) carrier
ULS	ultimate limit state
US	United States
USCG	United States Coast Guard
VDR	voyage data recording unit
VDU	visual display unit
VHF	very high frequency
VLCC	very large crude carrier
VR	velocity ratio
VTS	vessel traffic services
W (i)	representative of the ships displacement
W (ii)	winter loadline
W/L	waterline
WBT	water ballast tank
WMO	World Meteorological Organisation
Wp	waterplane area
w.p.s.	wires per strand
WPS	Welding Procedure Specification

Terminology and Definitions Associated with Marine Emergencies

Anchor handling vessel (AHV) – A high horse-powered vessel usually constructed with a wide, ample-spaced working deck, aft. They are frequently employed in offshore areas as a general-purpose work boat. They carry very long anchor cables in large lockers for their own use. These vessels are also used for transoceanic towing operations, usually having a high bollard pull (BP) capacity in excess of 130 tonnes.

Speciality towing with offshore structures and working in the salvage role are not uncommon. Such units are used extensively in the offshore industry for laying patterns of anchors for positioning offshore installations. Also employed for the recovery and deployment of anchors within the salvage sector.

Anchor warp – A wire hawser, sometimes combined with a heavy-duty fibre rope, which acts as an alternative to the anchor chain cable as fitted to conventional seagoing vessels. More often used on smaller or specialised craft where an all-chain cable would be considered inappropriate.

Arbitration – Defined as a method of settling disputes between two or more parties. Decisions from arbitration are usually binding on the parties concerned. The term is often common to 'charter parties'.

Arbitrator – A person designated to hear both sides of a dispute. The person is very often a Queens Council and in the case of salvage, such a person is likely to be practising at the Admiralty Bar under English Law relating to civil claims of salvage.

In determining any salvage award, account would have to be taken of the value of the ship, its cargo and freight at risk. Assessment would also be made of the dangers and difficulty in establishing salvage.

Archaeological salvage – A type of salvage for the recovery of either cargo or artefacts usually submerged and may involve the use of scuba diving. This type of salvage has evolved, with governments wishing to preserve wrecks and involve themselves in contracts to effect recovery of property of value and interest.

Backstays – An additional feature rigged to a mast or Samson Post structure to provide additional support when an attached derrick is expected to make a heavy lift.

Beaching – A term used to describe the deliberate action of running the vessel into shoals to take the ground. It is usually carried out to prevent a total constructive loss from the possibility of the vessel sinking in deep water. It would generally be expected that, following repairs, the vessel could be re-floated at a more favourable time in the future. An alternative is to break the vessel up in controlled conditions. The act of beaching would be cause for a declaration of 'general average' (GA).

Bimco Towcon – A widely used contract for sea towing. This towing contract was first introduced in 1985. It was drawn up by the Baltic and International Maritime Council (BIMCO), the European Tugowner's Association (ETA) and the International Salvage Union (ISU). This type of contract tends to incorporate 'Standard towing Conditions', and provides detailed and explicit conditions affecting both parties to the towing operation.

Bitter end – The opposing end of the anchor cable which is secured to the vessel in the region of the cable locker.

Bollard pull (BP) – A measured pulling capacity associated with the towing power of a tug. It is an influence on the charter towing rates when a tug is hired for a charter. The greater the BP, the higher the towing rate charged. It is defined as the amount of force, expressed in tonnes, that a tug can exert under given conditions.

Broken stowage – Considered to be that space contained between cargo parcels that remains unfilled.

Broker (insurance) – A third party who acts between the client who wishes to insure his operation and the underwriters who offer to take the risk on. The 'broker' acts to advise the client on an appropriate level of policy. The broker is informed by the Warranty Survey Company of changes to the operation or additional risks being incurred.

Bulk density – The weight of solids, air and water per unit volume. It includes the moisture of cargo and the voids, whether filled with air or water.

Bull wire – A single wire often used in conjunction with a 'lead block', rigged to move a load sideways off the line of plumb, to an acceptable position.

Cargo salvage – An occasion when a vessel is so badly damaged that it cannot be saved or the hull cannot be saved economically, but the cargo or part of the cargo can.

Cargo shift – A generic term used to describe an unwanted movement of the ship's cargo. It is usually experienced in bad weather where the vessel experiences violent motions in pitching or excessive rolling. The ramifications could affect the positive stability of the vessel, causing the ship to develop a list or even go to an unstable position. Avoidance of the problem is generally achieved by ensuring that the cargo is correctly and adequately secured after loading, before putting the ship to sea.

Carpenter's stopper – A heavy duty stopper employed to hold steel wire ropes (SWRs), used within the salvage industry.

Cement box – A temporary repair method applied to minor leaks about the ship's hull. It is established by the construction of a box in steel or timber around the area of the leak and fitted with a drain. It is then filled with cement and allowed to dry. The drain is led to a bilge compartment which can be conveniently pumped out.

Certificate of approval (C of A) – A certificate issued by a competent third party (Warranty Survey Company) in order for the insurance policy to become valid. In the event that the insured party does not abide by the conditions of the C of A, then the contract of insurance may become invalid. **NB.** Some issuing authorities use an alternative for the C of A: a certificate of transportation, a towage certificate.

Class surveyor – That representative of the 'Classification Society' that inspects and ensures that the vessel or structure remains compliant with the rules of the society and remains a 'classed' vessel or structure for the purpose of insurance and commercial reasons.

Cofferdam – An isolation space positioned between two adjacent compartments. Cofferdams are common to tanker vessel construction, serving as a separation between the accommodation block and cargo compartments. It may be a void space and always treated as an 'enclosed space' where full safety precautions must be adhered to, prior to entry.

Collision patch – An improvised repair to a breach in the ship's hull. A generic term for a temporary repair to the breach. A collision patch can be applied by the crew or a salvage team as appropriate to provide a quick seal to a damaged area of a ship's hull. Salvage teams may employ 'hook bolts' to secure the patch over a damaged area.

Common law salvage – Sometimes referred to as 'pure salvage', where property is recovered and no agreement or contract is drawn up. An example of such is where a ship recovers floating cargo from the sea and a claim for salvage is made on the owners of the property. Additional examples can be when property is found and recovered from the shoreline or from shallows, or by a diver making a recovery from the sea bed. A claim for salvage must then be registered against the previous owner of the property.

Composite towline – A towline which is established by coupling the ship's anchor cable to the towing spring of a towing vessel. The ship's anchor can be hung off aboard the vessel or left secured to the cable, to provide an increased 'catena' to the towline.

Contractual salvage – Salvage in which a salvage agreement, like 'Lloyd's Open Form' is drawn up between two parties as with the shipowners or between two ship's Masters, for the salvage of a ship and her cargo.

Cradle – A lifting base, usually manufactured in wood or steel or a combination of both, employed to accept and support a heavy load in transfer. It would normally be employed with heavy-duty lifting strops or slings from each corner.

Cutting operations – Several incidents have caused wrecks to be cut into removable sections, including the *Tricolor* and the *Riverdance*. The method of cutting is usually carried out by a 'diamond tipped cutting wire cable'. A sawing action is achieved by passing the cutting cable under the casualty, between the crane or sheer legs positioned either side.

Deadweight tonnage (dwt) – That difference between the ship's loaded and light displacement tonnages. Namely a measure of the cargo she can carry inclusive of water, fuel and stores.

Derelict – A vessel remaining afloat but in an abandoned condition. The first person to take possession of any derelict has the absolute right of control.

Displacement tonnage – The actual weight of a ship measured by the volume of water displaced, expressed in long tons.

Dynamometer – A load test instrument, either mechanical or in the form of an electric load cell, providing a numerical read-out on the load being handled.

Emergency steering – A generic term which expresses an additional/secondary means of steering in the event of failure of a main steering system. Not be confused with 'jury steering'.

Emergency towing arrangement – A requirement under SOLAS Chapter II, Regulation 1/3–4 requires all ships over 500 gross registered tonnes (grt) to be fitted with means of deploying emergency towing gear in a controlled manner, both forward and aft of the vessel's structure. Effective from 1 January 2010.

Even keel – A description of a vessel which is not listed either to port or to starboard. Even keel boats are easier to steer and handle than those which are hampered by an adverse list.

Fire monitor – Many vessels, especially tankers and salvage tugs, carry fixed water and foam fire monitors. They are capable of projecting pressurised water jets to a considerable range and height. They can be double functional for water or foam. Operating ranges vary depending on inlet pressure, but operation to 150 m would not be unusual (see page 84).

Floating Dock – A moveable dry-docking system for ships. They can be transported, usually by tugs, to any alternative waterborne position. The dock itself is a tank system which can be submerged and listed to allow damaged vessels to be docked even at an angle of heel. They are more frequently moored alongside a shipyard complex to handle any docking overspill, so ensuring continued work for the shipyard.

Flotsam – Goods which have been cast or lost overboard which are recoverable by reason of them remaining afloat.

Flow moisture point – That percentage of moisture content of a cargo when a flow state develops.

Free surface moments – Partially filled tanks found in many salvage incidents are frequently experienced, either totally flooded or breached. In such a case the positive stability of the casualty could be directly affected due to the uncontrolled movement of the liquid so contained in the tank, namely the 'free surface' movement of said liquid.

The virtual loss of stability can be ascertained by calculation using the following equation:

$$\frac{I}{V} \times \frac{d_1}{d_2} \times \frac{I}{n^2}$$

where: V = the vessel's volume of displacement, d_1 is the density of the liquid in the tank, d_2 is the density in which the vessel is afloat, n is the number of compartments into which the tank is subdivided and I is the second moment of area of the free surface. I can be calculated for a rectangle by:

$$I = L \times \frac{B^3}{12}$$

where: L and B represent length and breadth, respectively.

General average – An expression of joint financial responsibility of the shipowners, the cargo owners and of the Master and crew. Loss or damage of the ship or cargo by an Act of God, like a severe storm, would be shared by all parties. Where a loss occurred, say, through the fault of the Master alone, this would be termed 'particular average'. It is an insurance term and claims resulting from loss are usually settled by an 'average adjuster'.

Girding (or Girting) – An accidental and dangerous force pulling in the broadside direction on a towing line to the tug from the parent vessel. This force could be powerful enough to cause the tug to capsize. The tug very often employs a 'gog rope' over the towline in order to relieve tension in the towline to avoid the capsize.

Girdling – Increasing the stability of a vessel by increasing its beam at a position of the waterline. Often employed by heavy-lift vessels to increase the waterplane area to provide improved stability conditions when making an off-centre lift. Also employed during the salvage of the *Costa Concordia* in 2013.

Gog (fixed) – A strengthened lead for the guidance of the towline. A steel structure that provides a directional lead for the towline over the deck of the towing vessel. It may be used in conjunction with a 'variable gog'.

Gog (rope/wire) – A controlling/restraining element to the towline found in a towing operation. It is often directly connected to a gog winch or capstan to increase or decrease the weight and effective turning moment on a tug's towline. May be in the form of a short pennant to ease positioning of the towline and the influence of the main gog wire. **NB.** Also widely employed on-board anchor handling vessels.

Ground tackle – May consist of chain cable and anchor(s) deliberately laid to secure a vessel against unwanted movement. Frequently used in salvage operations where a stranded vessel is in danger of moving with the tide or weather (see also kedging).

Grounding – A term used to describe any contact of the ship's hull and the sea bottom, either accidentally or intentionally. Also known as stranding.

Guest-warp – A rope or line passed from the bow region to aft to hang just above the waterline, to provide convenience for boat crews to hold on to when coming alongside.

Heave to – A term that describes the action of holding the vessel in one position, usually against bad weather conditions. The ship's head is turned to meet the wind direction, while the engine is kept at reduced revolutions to maintain a holding speed over the ground.

Holding ground – The description given to the sea bed as to its anchor-holding properties. Mud or clay are considered as good holding grounds, whereas ooze or rock are considered as bad holding grounds.

International Salvage Convention (1989) – IMO convention on marine salvage specific to the needs of all parties concerned, regarding the activities of lawful salvage.

International Safety Management (ISM code) – A safety culture for shore-based and serving marine personnel within the marine industry. A system established to ideally provide safer ships and cleaner seas within the maritime environment.

Incompatible materials – Those materials which may react dangerously when mixed and are subject to recommendations for segregation in stowage.

Jetsam – Goods which have been lost or cast overboard from a ship which are recoverable through being either washed ashore or by remaining in relatively shallow water.

Jury steering – An improvised method of steering a vessel when designated systems to control the rudder have failed. A 'jury rudder' is one which is constructed of unrelated equipment like drag weights being deployed on either side of the ship to provide a turning effect.

Kedge anchor – An additional anchor, usually carried at the stern of the vessel. Employed for laying well astern of the vessel in the event that the ship has run aground. It would be used for the practice of 'kedging', i.e. pulling the vessel astern off the shoal.

Kilindo rope – A multi-strand rope having a non-rotating property often employed for crane wires. (Sometimes referred to as 'wirex'.)

Lagan – Goods cast overboard which are buoyed so as to be recovered at a later time.

Life salvage – Previously the salvor was not paid for saving human life. However, the law has now changed and where life is saved, as well as property, then 'life salvage' can be paid. Where salvage takes place to a British ship or a foreign ship inside British waters and the property salved is insufficient to pay the reward, then a discretionary payment may be made out of public funds by the Marine Authority.

Lifting beam – A long steel beam usually constructed as an 'H' section, employed to spread the weight of a long or awkwardly shaped load when being lifted.

Lloyd's Agency – An agency that manages marine insurance in three key areas:

1 A Lloyd's Agency Network, which operates with 330 Lloyd's agents operating worldwide with a similar number of subagents, to provide marine surveying and claims adjustment service for the global insurance industry.
2 A certificate office which produces certificated evidence of cover of marine insurance, in both paper and electronic formats.
3 A Salvage Arbitration Branch, which conducts the administration of Lloyd's Open Form of Salvage Agreement.

Lloyd's Open Form – A generic term employed widely throughout the maritime industry to describe the Lloyd's Standard form of salvage agreement. This agreement is now almost universally employed and is based on the principle of 'no cure – no pay', where the remuneration to the salvor is paid from the value of the property saved. It is a contract made between the party in peril (the shipowner and their representatives) and the salvor. It is an open contract where no monetary value is stipulated. The subsequent amount of any award would be determined by an arbitrator.

Load density plan – A ship's plan which tends to indicate the deck load capacity of cargo space areas. Used to ensure overloading of a deck area does not take place.

Lost buoyancy – The term given to a space on the vessel which has been breached and become flooded with water. For all intents and purposes, this compartment is no longer supporting the vessel and the added water inside the compartment becomes added weight.

Luffing – A term which denotes the movement of a crane jib or derrick boom to move up or down: luff up, luff down.

Lutine bell – A ship's bell salvaged from the ship *Lutine* which sank in 1799. It currently hangs in a room at Lloyd's of London and is traditionally rung prior to making an important announcement like a disaster at sea.

Mousing – An operation carried out to a shackle pin or the jaw of a hook to prevent accidental loss of a secure holding. When a shackle bolt is 'moused' seizing wire is employed between the bolt and the body of the shackle to prevent unforeseen movement of the bolt. Where a hook is 'moused' small stuff (cordage) is used under the bill of the hook and around the neck of the hook to prevent any load jumping off the bill. Larger hooks are often fitted with a spring-loaded 'mousing tongue' to achieve the same objective.

Not under command – The description given to a vessel which through some exceptional circumstance is unable to manoeuvre as per the regulations for the Prevention of Collision at Sea.

Oceangoing tugs – The largest of the tug categories and defined by size. Frequently employed in salvage towing operations, having a capacity range from 4,000 hp to 22,000 hp. They tend to have the higher BP capacity of all the tug groups.

Oil skimmer – A specially designed vessel to clean and recover oil slicks spilled to the surface of the water.

Outreach – A term used with crane operations to express the maximum working distance of a lift operation, measured from the pivot position of crane or derrick. The distance is directly influenced by the length and angle of the crane jib (boom).

Overhauling – A term used to describe the correct movement of a block and tackle, (lifting purchase) arrangement. The term indicates that all sheaves and wire parts are moving freely without restriction.

Parbuckle – An old system employed for raising and lowering loads by use of two or more ropes around an object. One end of the ropes is secure and the load is moved in the bight, by hauling or easing from the other end. The system has been used in the salvage industry to right a capsized vessel when laying on her side, e.g. the *Herald of Free Enterprise* and the *Costa Concordia*.

Particular average – A loss of a ship or cargo as caused by an Act of God, it can be related to a ship stranding, in collision or on fire (*see also* general average).

Permeability – A measure of the ship's space that if flooded can be occupied by water. The nature of cargo as carried by the vessel will generate a different permeability factor. Expressed as a percentage.

Permit to work – A procedural safety checking system which operates within the maritime industry (and other industries), wherever electrical work, high levels, hot work or enclosed space entry is required to function.

Pollution boom – A barrier or floating boom which when deployed around a stranded vessel is meant to act as a restrictive measure to prevent the spread of oil or similar pollutants. Frequently referred to as a barrier for restricting oil spills, otherwise known as boom equipment.

Proof load – That tonnage value that a lifting appliance, e.g. crane or derrick, is tested to. The proof load equates to the safe working load plus an additional percentage.

Ramshorn hook – A heavy-duty, double lifting hook capable of accepting slings either side. Extensively employed in heavy-lift work within the salvage sector of the industry.

Receiver of wreck – A person of a district who is appointed on behalf of the Treasury of the United Kingdom. The Maritime Coastguard Agency would normally take charge of any incident and act as the receiver. However, they may appoint a customs officer or a representative from the Inland Revenue to act as 'receiver of wreck', or any other person deemed suitable.

Release note – A signed note by the Master of the casualty on the completion of salvage services. Usually administered when the ship is safely afloat at an agreed location and the Master accepts re-delivery and satisfaction of the salvage services given.

Risk assessment – A detailed, investigative report conducted prior to the commencement of practical operations. It should contain such elements as the manpower involved, prevailing weather conditions, equipment reliability, emergency communications and any other elements that could generate an intolerable risk as opposed to an acceptable risk.

Rope gauge – A handy measuring device used to ascertain the diametric size of a rope or wire.

Safe working load – An acceptable working tonnage used for a weight-bearing item of equipment. The marine industry uses a factor of one-sixth (1/6) of the breaking strength to establish the safe working value.

Salvage (maritime) – A term which describes the actions of a third party where property is saved from a peril of the sea. It can also be represented to define the money paid to a 'salvor' as remuneration for his/her services. When related to a marine insurance policy a 'salvage award', or part thereof, which is determined as recoverable under the policy. **NB.** In non-marine practice the term salve is used to represent goods saved from a land-based fire.

Salvage agreement – An express agreement entered into by parties to a salvage operation. Lloyd's Open Form of Salvage is a widely used standard agreement. Such agreements are usually based on a 'no cure – no pay' basis and provide either a fixed sum or a derived sum, settled by arbitration.

Salvage Association – The world's leading organisation of marine casualty and investigation surveyors. It was established in 1856 in London to serve the interests of underwriters, shipping and cargo activity.

Salvage craft – A generic term which describes a variety of designated salvage vessels. They are often equipped with heavy-duty lifting devices like sheer legs or heavy-duty cranes with associated deck and barge type carriage space.

Salvage lien – A maritime lien on property salved, which gives the right to arrest the property even if it has changed hands. The purpose of the lien is to allow the property to be sold in order to meet legitimate claims.

Salvage operation – Any act or activity undertaken to assist a vessel or any other property in danger in navigable waters or any waters whatsoever (definition by the International Maritime Organisation).

Salvor – The name given to that person claiming and receiving salvage for rendering services to save the vessel and cargo, or any part thereof, from an impending peril, or for having recovered the same after actual loss.

Seaworthiness – A vessel is deemed seaworthy when reasonably fit in all respects to encounter the ordinary perils of the sea. It is also an implied warranty clause when in the format of an ordinary marine insurance policy.

Secondary pollution – Second-stage pollution which may occur from an initial incident site. A typical example of secondary pollution was experienced with the *MSC Napoli* container vessels where some of the containers when removed were found to be leaking and contaminated with oil. These cargo parcels had to be separately held in a safe area for cleaning to avoid secondary pollution occurring.

Shackle of cable – A measure of anchor chain cable set at 15 fathoms, 90 feet or 27.5 metres. Where ground tackle is employed a number of 'shackles' would be employed.

Sheer legs – An 'A' frame type lifting structure mounted to a floating pontoon barge. They are usually self-propelled and fitted with an accommodation block. Although not originally designed for salvage work they have become well known to be active in salvage operations, with a lifting capacity to 3,000 tons. Although they act in the way of a crane, sheer legs can be fitted with an extension fly-jib to its main lifting jib, so providing an extending lift capability.

Ship-to-ship transfers – Salvage methods vary considerably, but many operations include reducing a casualty's load by the transfer of cargo to another vessel, so lightening the initial load. Such ship-to-ship transfer can be a double achievement for the salvor: (1) cargoes can be recovered and salved, while (2) the ship itself can be lightened and caused to re-float.

Shore – A term used to describe a support, given to decks, bulkheads or cargo. They are usually in timber but may be in the form of a metal adjustable stanchion, depending on usage.

Skeg – A centre line construction extending the aft-most part of the keel. It is a common structure to modern twin-screw ships. In some cases it is large enough to accommodate stern thrusters. When a vessel trims by the stern, as most do, the lower part of the skeg could well be the lowest depth of the vessel and as such is the most vulnerable to grounding.

Sounding – A method of measuring the liquid amount inside a tank. The term is also used to ascertain the depth overside from the ship.

Stowage factor – The volume occupied by unit weight of cargo. It is an expression which indicates the cubic capacity that 1 ton of cargo will require. It should be noted that this is not the actual cubic capacity of 1 t of cargo as the stowage factor takes account of the design and shape of the cargo package and special stowage arrangements required, e.g. use of dunnage.

Territorial waters – Those waters which lie adjacent to a country's shoreline and over which that country claims jurisdiction.

Towing point – A variety of vessels in the maritime sectors are often engaged in towing and the vehicles, like barges, offshore rigs, floating cranes, etc., are usually fitted with designated towing points. Tanker vessels are generally fitted with emergency towing devices in the form of 'Smit brackets' or similar arrangements. Other towing points are specifically designed to accommodate a chain securing or may employ double bollard sets where a purpose-built towing point is not available. Many arrangements actively engage a chain bridle, fixed between towing points to spread the load of the tow.

Towing spring – An element of the towline which has a high breaking load but also has a stretch ability to act as a shock absorber in the event that the towline is 'snatched' and takes an abnormal load to that expected. The towing spring is usually secured to the end of the towing wire or directly onto a 'bridle', if employed.

Towline – That line which is secured between the towing vessel and the vessel being towed. The composition, length and cross-section of the towlines are variables and reflect the overall weight that the towing operation will expect to experience in differing weather and sea conditions. Weight for weight, the world's strongest material is UHMPE, and permits size-for-size replacement of SWRs. It is lighter and easier to handle than alternative towlines, which require equivalent strength.

Trimming – A manual or mechanically achieved adjustment to the surface level of a bulk cargo as stowed in a cargo space. This is usually carried out to provide greater stability and prevent cargo from 'shifting'.

Tug approval survey – An inspection survey of the towing vessel conducted by the warranty surveyor. The purpose of this survey, usually conducted by a checklist, is to ensure that the tug is fit for purpose and without defects.

Ullage – That measurement above the sounding, taken from the surface of liquid inside a tank to the underside of the tank top.

Underwater salvage (wet salvage) – A common method of salvage of a variety of select products which employs the numerous advances in technology, inclusive of remote-operated vehicles (ROVs), dynamically positioned vessels, divers and/or diving bells, together with the use of underwater cameras.

Underway – A vessel is said to be underway when not at anchor, made fast to the shore or aground.

Underwriter – A person who is authorised to carry out insurance business on behalf of the insurance company or a Lloyd's syndicate. He or she will take an underwriter's commission and will find other applicants through sub-underwriters to spread the risk of the insured commitment.

Underwriters – Those people or organisations that provide insurance cover. This is achieved in two ways: (1) by the wealth of an insurance company to meet subsequent claims, or (2) Lloyd's syndicate members providing a capital base to meet subsequent claims.

Warranty survey company – A specialised company providing consultancy services to the offshore and marine industries. Depending on the size of the practice it will generally contain expert personnel in any or all of the following disciplines: marine engineering, meteorology, seamanship, naval architecture, hydrography and marine law.

Warranty surveyor – A designated person from a warranty survey company.
He or she will be supplied with a 'job package' inclusive of contact details for the client, together with instructions as to the inspection criteria and guidelines relevant to the operation on hand.

Wreck – A ship or parts of a ship that remains and is no longer serviceable or capable of navigation, found on the sea bed or the shoreline. It may or may not have accompanying cargo.

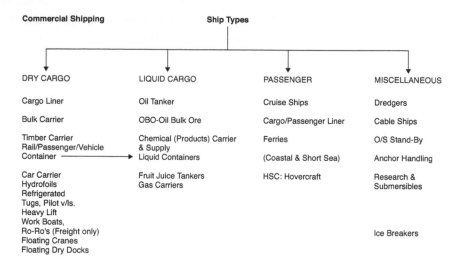

Commercial Shipping   **Ship Types**

DRY CARGO	LIQUID CARGO	PASSENGER	MISCELLANEOUS
Cargo Liner	Oil Tanker	Cruise Ships	Dredgers
Bulk Carrier	OBO-Oil Bulk Ore	Cargo/Passenger Liner	Cable Ships
Timber Carrier Rail/Passenger/Vehicle Container	Chemical (Products) Carrier & Supply Liquid Containers	Ferries (Coastal & Short Sea)	O/S Stand-By Anchor Handling
Car Carrier Hydrofoils Refrigerated Tugs, Pilot v/ls. Heavy Lift Work Boats, Ro-Ro's (Freight only) Floating Cranes Floating Dry Docks	Fruit Juice Tankers Gas Carriers	HSC: Hovercraft	Research & Submersibles Ice Breakers

Specialist: Fishing v/ls, Factory ships, Auxiliaries for warships, Light vessels, Sail Training, Harbour craft, Tenders & bunker barges, Oil Recovery vessels, Emergency Support and Salvage craft.

Figure 0.2 Ship types and carriers of the Mercantile Marine.

1

Collision (Ship to Ship) at Sea

Introduction

With any collision at sea the number of variables will not only influence the outcome but generally means no two collisions are ever the same. Each collision will be unique because of the position of contact, the weather prevailing at the time, the geography, a loaded or light condition, Masters' experience, day- or night-time scenario, etc. So many factors will differ that no one, least of all this author, could hope to provide an answer to every situation. The very best that can be produced is to develop a general format that would be acceptable for the typical, average incident.

Clearly, the avoidance of collision in the first place is the obvious way to go, on the premise it is better to prevent than to cure the after effects. However, we do not live in a perfect world and accidents do occur. The point of impact in a collision and the subsequent damage will differ accordingly. Subsequently, the immediate and long-term corrective actions will differ in accord with each scenario. A noticeable example can be readily seen where two similar collisions take place. One vessel is struck below the water line and would probably require pumps to be activated on the affected area, whereas another vessel is struck above the water line and doesn't lose water-tight integrity and has no need to put pumps into operation.

The law of providence could have major ramifications for casualties so involved in collision incidents. What cannot be left to chance are the legal aspects surrounding a collision, or the medical treatment required by any casualties so involved. Masters and senior officers receive little or no experience in dealing with a real-time casualty incident until they find themselves in the thick of it. Potential background training can have a limited effect, but a strong belief in the first principle of the Safety of Life at Sea is by far a greater motivator to do what is right and necessary.

Collision: Immediate Effects

Reactions following any collision at sea are bound to generate a certain level of shock among personnel on board any vessel so involved. This sudden shock experience can expect to last an indefinite period of time. The ramifications of not acting positively as soon as practical after this initial period of shock are not worth contemplating. In other words: get over it and let the training kick in.

Certain ranks within the shipping industry have usually had a degree of emergency training and hopefully will react positively and practically. Actions being based on the first principle of 'The Safety of Life at Sea' are paramount. If the position of the Master is considered, he has a legal obligation to stand by to render assistance to the other vessel. This is all very well, but could a man or woman think only of a third party's needs, in isolation to his own ship and own crew's safety?

Any actions by the Master or Officer in Charge of either vessel can only be made from a position of strength. Therefore, unless he or she wants to escalate the situation, certain basic needs have to be fulfilled quickly. An immediate requirement for the person in charge is to take the 'conn' of the vessel and establish a command chain. Sounding the general alarm as soon as possible if it has not already been initiated could be seen as the most immediate of activities. However, it should be realised that no single individual can expect to do everything himself; he must delegate activities to realise best effect.

A history of drills during routine voyages can prepare officers and crew members for that unexpected emergency incident. If personnel know their stations, then the chain of command can expect to permeate through any catastrophe. Activities need to be prioritised; there is no point in sending a distress message before obtaining a position or gaining knowledge of the immediate problem(s).

A series of activities should take place, probably starting with ship's officers reporting directly to the bridge following impact damage. The Master would expect to order his Chief Officer to carry out an initial 'damage assessment', while the Second Officer (Navigator) would probably take over as the Officer of the Watch and obtain the ship's position. (Different companies/ships employ different ranks in differing roles.)

Personnel could be expected to take up the duties of helmsman and lookouts, while a third mate could be designated as communications officer. Each incident would expect to generate exceptional activities, over and above normal routine. Certain activities on certain ships can be coordinated quickly, like the closing of watertight and fire doors. Or, for example, placing engines on 'stand-by' for immediate readiness where the ship is not fitted with bridge control capability.

Correct interpretation of data will allow critical activity to reduce casualties and loss of life. Incoming information should fit into an acceptable framework which takes into account all eventualities.

> **NB.** This assumes Masters and ship's officers are alive and capable of conducting emergency operations.

An example collision framework could include any or all of the following immediate actions:

For the role of Master (after collision impact at sea)

1 Move immediately to the navigation bridge and take the conn.
2 Stop the ship's own engines if underway and making way, depending on the position of the collision and how the vessels have struck. A few revolutions on engines could reduce the permeability by keeping the bow plugged into the damaged area.
3 Sound the general emergency alarm if not already activated.
4 Order a roll call and check the ship's complement for casualties.
5 Close all watertight doors.
6 Close all fire doors.
7 Order the engineers to stand by and go to an alert status in the machinery space.
8 Obtain the ship's position in latitude and longitude by any reliable means.
9 Turn the ship's 'deck lights' on and display the vessel's 'not under command' (NUC) lights.
10 Designate an immediate Communications Officer.
11 Order the Chief Officer to obtain an interim 'damage assessment'.
12 Order the muster of 'damage control parties'.
13 Activate deck parties to turn out lifeboats and a rescue boat.
14 Bring a bridge team together to include lookouts and helmsman.
15 Order a local weather forecast to be obtained as soon as practical.

NB. Under GMDSS, a dedicated Communications Officer is appointed. Where officers are limited in numbers this may not be a practical option to isolate an officer to this duty alone – handling communications to the detriment of all else.

Peace of mind will certainly not occur until the ship's Chief Officer provides the outcome from his initial damage assessment. Even then such a report may bring additional problems. Any Master will have the role of communications immediately following a collision, but he needs to know the subject matter of expected communications. We have already seen the catastrophic outcomes of mixed communications in the incident of the *Costa Concordia*.

Miscellaneous 1. Where a vessel has incurred collision in such a manner as to be left embedded into the other vessel, then it may be appropriate to leave the vessels in contact rather than separate the two ships. This could be achieved by maintaining a few revolutions on the engines of the striking vessel.

The reason not to separate is to retain one ship acting as a plug to the other and so reduce the permeability factor. Effectively this action could stop excessive flooding to the damaged vessel.

Another reason to stay in close contact could also be if one of the two vessels involved is a tanker-type. Clearly a full oil tanker, if damaged, is unlikely to sink, on the basis that oil floats on water. However, tearing metal hulls, when separating the two ships, could cause ignition to the fumes and gases being released from tanker cargoes. Collision with any ship is bad, but with a tanker the real danger of fire as a result of lack of thought could be even more threatening.

Miscellaneous 2. Where a vessel has been struck below the water line and the watertight integrity of the hull has been breached, flooding will take place. A prudent Master would without doubt direct the engineers to activate the pumps to any effected area.

The pumps would probably not be able to handle a continued, major ingress of water, but the use of pumps could buy valuable time before the ship loses buoyancy.

The circumstances of each incident will influence subsequent actions. For example, if a vessel is struck in one of its sides, at a water line position, then the action of adding ballast to the opposite side may be sufficient to raise the damaged area above the waterline by deliberately giving the ship an adverse list, so reducing water ingress.

Impact Damage

Figure 1.1 An offshore all-purpose stand-by vessel is seen in dock to effect repairs on its starboard side. The damage is on and about the waterline, into the topsides and 'boot topping' paintwork lines. The scale of the damage would suggest that a collision patch could be appropriate immediately after occurrence. Long-term, the watertight integrity of the hull is lost and the ship is inevitably docked to undergo repairs.

Legal Actions of the Master (Under the Merchant Shipping Act)

The legal obligations of each Master or Officer in Charge involved in a collision is to '*Stand by the other vessel to render him assistance*'. This is not an easy task if your own vessel is badly damaged and the safety of your own crew is in question.

M/S 1995 Section 92, states:

> In every case of collision between two ships, it shall be the duty of the Master of each ship, if and so far as he can do so without danger to his own ship crew and passengers (if any) –
>
> *a) to render to the other ship, its master, crew and passengers (if any) such assistance as may be necessary to save them from any danger caused by the collision, and to stay by the other ship until he has ascertained that it has no need of further assistance.*

This activity could clearly be directly influenced by the Master's second legal obligation, '*To establish communications and exchange information*'. Although it is specified what information must be exchanged, it stands to reason that additional items related to the collision will also probably be discussed, if time allows:

> *b) To give to the master of the other ship the name of his own ship and also the names of the ports from which it comes and to which it is bound.*

NB. Communication and exchange of information would not seemingly be a difficult task, provided both parties speak a common language. Even though 'English' is the international language of the sea, there are no guarantees that both parties to a collision have this fluent commonality.

It should be borne in mind that the Master has four legal obligations in the event of collision:

1 Stand by to render assistance to the other vessel.
2 Exchange information with the Master or Officer in Charge of the other vessel.
3 Cause an entry of the incident to be entered in the official log book.
4 Report the incident to the MAIB of the Marine Authority as soon as practicable after the incident (Ref., MGN 289).

Immediate damage assessment to permit external communications

On:
- watertight integrity of the hull
- engine room wet or dry
- casualty report
- indications of pollution.

Commonsense should prevail when senior officers complete necessary paperwork. There is a time and a place, which tends not to be when casualties need immediate attention. It is normal practice for Masters to make a note of timings and occurrences in a pocket note book before making direct entries into log books. This tends to ensure greater accuracy when making final reports.

General Actions Following Initial Response to a Collision

Assuming both vessels are standing by each other, the Master of each vessel would probably be looking to obtain more detail on the initial damage assessment. This could be enhanced by all or any of the following:

- Obtaining a full set of internal tank soundings to confirm broached tanks and intact tanks of one's own vessel.
- A full stability assessment with the existing/worsening conditions. Assess continued positive stability even after subsequent flooding.
- A detailed list of casualties, if any, together with an assessment of injuries and injury-related deaths. There should also be a report of any persons lost. (Life-threatening injuries could determine the grade of communications, whether to go to a Class 1 Priority MAYDAY or to a grade 2 URGENCY.)
- Checks to be made on essential machinery and power supplies, the results of which could dictate the use of the emergency generator.
- Checks to be made on essential navigation equipment to ensure that course, speed and position monitoring can all still be achieved. (Even if it becomes necessary to get underway under NUC signals.)
- Inspect the area for fire risk or chemical incursion, both from one's own vessel and from the other vessel. It would be essential to obtain a weather forecast and determine the local wind direction in most cases. This is especially relevant in the case of toxics being blown down onto one's own ship from the other vessel.
- The initial position of the collision should be confirmed as soon as possible after the impact. Subsequent monitoring of that position should be regularly ascertained for communication purposes, especially in a sea area affected by tides or currents.
- Communications with the other vessel should reveal the possibility of any associated dangers from: explosion, toxics or fire. One must assume that each Master would wish to know the nature of each other's cargoes, which may influence the action of both ships involved.

- One's own crew would be expected to be ordered to respective damage control parties. These may vary from maintaining boundary cooling, to actions to reduce the ingress of water. Other personnel could be delegated to operating as stretcher parties/first-aiders, depending on circumstances.
- Depending on the severity of the situation, a Master must consider the possibility of abandoning the vessel. This would result in the need to turn out survival craft to the embarkation deck.
- The Master would also investigate navigation towards a port of refuge. This would clearly depend on the ability of the ship to get underway. Depending on circumstances and the time element, an alternative to making the port of refuge may be the possibility of beaching the vessel.
- The option to beach the vessel is only possible, of course, if a beach scenario is within range and the capabilities of the damaged vessel. Beaching is carried out generally to prevent a total constructive loss (see Chapter 2 for extended detail on beaching/grounding).
- Where delaying tactics and/or damage control efforts are not showing positive outcomes, the Master must transmit either a MAYDAY or URGENCY signal. In the case of a passenger vessel the decision is made for the Master – it is clearly a MAYDAY.
- The option to send an URGENCY signal must always be considered as prudent on the basis that this can always be upgraded to a MAYDAY.
  - It should be noted that a MAYDAY signal can be downgraded to an URGENCY situation.
  - In any event, Masters must cancel a MAYDAY signal once the situation is resolved.
- If it is ascertained that the vessel can remain afloat despite damage sustained, then the option to take a towing offer may have to be considered. There should be a realisation that accepting a towline could leave the vessel and cargo liable to a claim for salvage (see Chapter 7 on marine salvage).

NB. Masters need to consider the ramifications of not taking a tow from a capable vessel against the immediate threat to his/her own crew and passengers.

Incorporation and Use of Checklists

Prior to the introduction of the International Safety Management (ISM) system, many tasks on board many vessels were covered by company standing orders and/or formalised check lists. ISM, once established, placed a greater emphasis on the use of checklists for many routine incidents as well as for specific emergency situations.

Officers in Charge were encouraged to employ the use of checklists as an everyday item even for routine tasks such as testing and checking bridge navigation equipment prior to leaving port, testing emergency steering gear or conducting emergency drills. The use of a checklist ensures that all items and actions are adequately covered, but at the same time it makes personnel more familiar with operational systems. The checklist is now also seen as another method of on-site training for individuals.

Specific incidents like collision or grounding are generally one-off scenarios and personnel cannot expect to be as familiar with the needs of such an irregular occurrence. A communication contact may easily be missed in an emergency, whereas if a checklist is used emergency contacts might not be forgotten in the anxiety of the moment.

It is a requirement that passenger ships carry an emergency SAR plan on the navigation bridge. It would follow that associated checklists for relevant activities and communications to conduct SAR operations would be attached or close to hand. A checklist, once completed, would remove the worry in an emergency situation, that everything that could be done has been done.

A well-prepared company will have checklists for virtually every occasion and for each specific shipboard task. Having said that, they should not be treated as 'dummy cards' and persons in authority must also think outside the box where different circumstances may not fit the order of the checklist. Specific criteria will normally incorporate a level of flexibility and should not be contained in a rigid format without taking prevailing conditions into account.

Even with the use of a checklist, systems are not infallible and should be employed with experience and an open mind. Our industry is influenced by many variables, not least the weather and geography. No checklist can have built-in accommodations for every eventuality and should always be used with an element of caution and in conjunction with the letter of the law.

Example scenarios usually covered by shipboard/company checklists include:

- collision with another vessel;
- collision with a fixed obstruction;
- running aground;
- operation of total flood CO_2 system;
- preparations when encountering heavy weather;
- search and rescue (SAR) operations;
- fire on board;
- helicopter operations;
- a death on board the ship at sea.

Communications Following a Collision

The decision to either transmit a MAYDAY or an URGENCY signal is one that is not always easy to define. By way of providing a guide, it is suggested that where a Master has death or injuries likely to result in death a MAYDAY signal is justified. If abandonment of the parent vessel is ordered and persons are entering a hazardous situation by taking to survival craft then, again, a MAYDAY signal would be justified. Based on the fact that the authorities would rather know sooner than later in order to initiate their response to reduce the subsequent loss of life, a MAYDAY signal or MAYDAY relay would be anticipated in every case where a passenger vessel is involved.

MAYDAY relay – Action taken to transmit a MAYDAY, where the vessel in distress is unable to despatch the signal by his own means. The MAYDAY relay being sent on behalf of a third party, e.g. where the other vessel has lost all power.

Figure 1.2 Example of GMDSS communications console to satisfy the requirements for all vessels over 300 grt on international voyages to carry specific satellite and radio communication equipment for the sending and receiving of distress alerts and maritime safety information (MSI), inclusive of navigational and meteorological warnings and forecasts.

Voyage Data Recorder (Black Box Recorder)

It is now a requirement under SOLAS Chapter V, Regulation 20, that all passenger ships and all vessels over 3,000 grt which are engaged on international voyages must be fitted with a voyage data recording unit (VDR).

The Maritime Safety Committee (MSC) has also agreed that VDRs are to be fitted to all existing roll-on–roll-off passenger (ro-pax) vessels and high speed craft already in operation.

The principle of marine 'black box technology' has come about because of the transport relationship with the aviation industry, which has been operating with black box technology on all passenger aircraft for many years. The monitoring of all principal elements within the mode of transport has shown itself indispensable in resolving aircraft accidents and subsequently improving long-term industrial safety.

The revised Chapter V of SOLAS has made carriage mandatory for certain types of vessels. The International Marine Organisation (IMO) has stipulated the data that VDRs are expected to record, which includes the following:

- date and time
- ship's position and speed
- course/heading
- bridge audio – one or more microphones situated on the navigation bridge to record conversations near the conning position and at relevant operational stations like radar, chart tables, communication consoles, etc.
- main alarms and PA systems
- engine orders and responses
- rudder orders and responses
- echo sounder recordings
- status of watertight and fire doors
- status of hull openings
- acceleration and hull stress levels (only required where a vessel is fitted with response monitoring equipment).

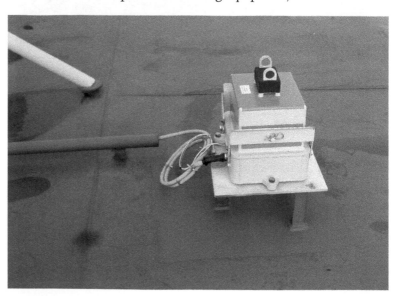

Figure 1.3 An example of a voyage data recorder (black box) seen fitted to the external decking of the 'Monkey Island' above the navigation bridge of the ro-ro vessel *Clipper Point*.

The information should be stored 24 hours a day over a seven-day period. It should be contained in a crash-proof box, painted orange and fitted with an acoustic device to aid recovery after an accident. The system will be a fully automatic, memory unit being constructed to be 'tamper free' and always watching, even when the vessel is tied up alongside. VDRs are expected to provide continuous operation for at least two hours following a power failure and are also alarm protected in the event of malfunction of any of the VDR's elements.

In the event of a marine accident or where the loss of a ship occurs, the marine authorities or the shipowners would expect to recover the unit and at least the last 12 hours of recorded data.

Note of Protest

The Master of a ship involved in a collision incident would also consider making a Note of Protest through a 'Proper Officer', i.e. the British Consul. Below is an example Note of Protest:

MARINE NOTE OF PROTEST

On this .... day of .... 20**...., personally appeared and presented himself before me .... (insert name of Master), Master of the .... (insert name of vessel) of .... (insert name of Port of Registry), official number .... and ...... net tons, which sailed from .... on or about the .... (insert date) in ballast/with cargo of .... (insert brief description of cargo if any) bound for .... (insert desti-nation port), and arrived at port of ....

.... on (insert port name and date) and declare that during the voyage having experienced a collision with the .... (insert name of colliding vessel) at about .... hrs on ... ... (insert date of collision) in .... (insert general location of collision, e.g. 'in Black Sea') and fearing damage to the vessel's hull, rigging, machinery and/or cargo, he hereby Notes his Protest against all claim, losses, damage etc., reserving the right to extend the same at time and place convenient.

Sworn and sign before me ...., Captain (insert name of Master)

At .... Notary Public.

Notes

_____

A similar Note of Protest would be applicable for other situations like 'Stranding', where the word 'collision' would be changed for 'stranding' or 'fire', whatever was relevant to the circumstances.

A Note of Protest would also be registered in all cases of general average (GA) being declared.

The Role of the Ship's Chief Officer (in the Aftermath of Collision)

The associated noise and movement of a collision would generally be expected to alert persons of an incident. In any event, the sounding of the general alarm would without doubt confirm the unexpected has occurred. Provided crew members have been practised with drill procedures, most would be expected to take up their emergency stations.

In the case of the Chief Officer it would be expected in the majority of cases for him/her to report to the navigation bridge. It must be anticipated that little certain information about the incident is as yet known. It is usually the Chief Officer who will be ordered by the Master to carry out an initial and immediate damage assessment.

The purpose of the Damage Assessment is twofold: first, to provide information on the accident to the Master, to allow him to make informed external communications. The second is to provide insight as to the extent of the damage and so give an indication of subsequent actions to provide a positive way forward.

The Initial Damage Assessment must provide detail on the following topic areas:

1 watertight integrity of the hull;
2 the condition and state of the machinery space;
3 a casualty report;
4 indications of any marine pollution.

The essential four topics will allow the Master to formulate the detail of his external communications, which may need to be despatched sooner rather than later.

The Chief Officer would expect to leave the Master on the 'conn' of the vessel and return to either damage control duties or preparations for going into an abandonment phase.

He would certainly order a full set of tank soundings as soon as is practical, so as to carry out an early stability assessment. The use of 'damage stability information' must be anticipated, especially where bilged compartments are present. The Chief Officer will also order the lifeboats and/or other survival craft to be turned out, ready to allow a speedy evacuation if this becomes necessary.

He or she would also direct operations towards the wellbeing of any casualties and work closely with any instructions and directions from the Master.

Depending on the nature and position of the impact and the developments from such a collision, the Chief Officer's orders to manufacture a 'collision mat' or move to effect temporary repairs may be needed. Delegation of duties to Petty Officers and senior men and engineers could be expected in order to reduce the effects of the collision escalating.

Collision Patch Construction

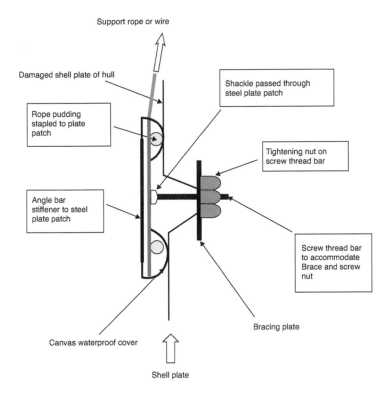

Figure 1.4 Collision patch construction.

Support rope or wire

Damaged shell plate of hull

Shackle passed through steel plate patch

Rope pudding stapled to plate patch

Tightening nut on screw thread bar

Angle bar stiffener to steel plate patch

Screw thread bar to accommodate Brace and screw nut

Canvas waterproof cover

Bracing plate

Shell plate

Collision Patch Materials

Not many vessels will carry designated collision patch materials. Therefore the majority of ships would be expected to improvise, should a collision patch become necessary. Customised equipment would probably include steel plate and a welding set, together with additional equipment in the way of steel angle bar or marine plywood.

Where designated equipment is not carried, improvisation could be the name of the game. Virtually all vessels would have bottom plates to the engine room. Such steel plate could act as a substitute. The average ship would carry waterproof plastic sheet or canvas. With these items the collision patch itself could be manufactured. Plates can be welded or bolted together to provide a covering patch to mask any hull damage. Such a plate would be fitted with a centre shackle and waterproofed.

> **NB.** Waterproofing could be achieved by providing a pudding around the perimeter of the plate/patch and covering the whole with waterproof sheet. The exposed shackle is positioned facing inboard, to the central area of the patch.

The task to position the patch over a damaged area would probably be challenging. If the assumed damage is on one side or the other, listing the vessel over to raise a damaged area over and above the waterline would lend to positioning any patch. Once constructed, the waterproofed plate could be lowered over the gunwale and allowed to slide down the ship's side to cover the damaged area from the outside.

Once the central shackle is seen in the centre of the damaged area a wire on the bight could be passed through the shackle and tensioned to bring the patch inward to the hull plate and achieve a compressed seal around the damaged area.

Such improvisation is not meant as a permanent seal, but could be managed by regularly pumping bilges until a 'port of refuge' could be realised in order to instigate more effective repairs.

Good seamanship is very often improvisation of materials that are available at the time.

Port of Refuge

In many incidents of collision or grounding, ships may find themselves in immediate peril. So much so, that the wellbeing of the ship and the continued voyage could lead to loss of life or property. When such an event occurs, Masters or Officers in Charge may be forced to seek out a 'Port of Refuge'.

The Port of Refuge is defined by the ship's need to deviate to another port other than the designated destination or when the ship has to return to the port from which she departed. The reason for such action must be genuine to fulfil insurance rights and other contracts and may include effecting necessary repairs after an incident.

The vessel could also seek the port of refuge to take bunkers, provided that when she left her last port she had an adequate reserve on board. An incident may have caused the loss of bunkers or it might have been found that the bunkers were contaminated in some way, where it becomes necessary to obtain suitable and adequate bunker oil.

In the event that a Master seeks a 'port of refuge' he/she must communicate with the following parties:

- shipowners – to advise of the selected port and the reasons for the deviation;
- the port authority – to request free pratique;
- the customs authority – to obtain clearance inwards;
- a 'Proper Officer' – to note protest;
- owners – to advise of safe arrival;
- underwriters or Lloyd's agent – to inform them of the accident in accordance with the 'Tender Clause'.

The ship's Master would need to ascertain the following details:

1 the costs of any damage to cargo;
2 the costs of discharging any cargo parcels;
3 the cost of fuel and stores necessary to effect repairs to the vessel and instigate required surveys;
4 damage surveys for hull and machinery and repair tenders for the same;
5 agreement with underwriters for repair tenders;
6 port costs, light dues, pilotage and berthing overheads;
7 costs of reloading cargo parcels, fuel and stores for onward voyage;
8 any storage and insurance costs incurred;
9 costs of any towage involved;
10 wages and maintenance costs of the Master, Officers and crew, and the ship's day-to-day upkeep;
11 where necessary to proceed to a second port of refuge, all costs involved;
12 any costs of transhipment of cargo to a second port.

Following repairs to the vessel, made to the satisfaction of the Classification Society, the surveyor would issue an 'Interim Certificate of Class'. In the event of the survey being conducted by a private surveyor, preferably one recommended by the underwriters, a Certificate of Seaworthiness would be issued. Either document would normally be acceptable to underwriters to allow the vessel to proceed on her voyage.

Port of Refuge and General Average

In some circumstances where a ship deviates to a port of refuge, a declaration of 'general average' may be made where the ship, cargo or freight are in dire peril. Other occasions involving a port of refuge, where general average may be declared are:

- where the vessel is experiencing fire, cargo shift, collision, grounding or leaking during a loaded voyage;
- where a vessel has to effect essential hull or machinery repairs;
- when necessary to take a tow after machinery failure.

NB. The intentional act of beaching a leaking ship, when loaded with cargo (i.e. voluntary stranding) to prevent the vessel from foundering is also generally allowed for in a 'general average' claim.

General average is a marine insurance term (under the York–Antwerp Rules) which is used for adjustment of loss when cargo on board a vessel, belonging to one or more owners, has been sacrificed for the safety of all those party to the venture.

The actual loss is shared by all those who have shipped cargo in the vessel. The claim for 'general average' to be made is when the loss has been voluntary, not accidental or caused by any fault on the part of the owner who is claiming the general average. The deliberate action taken, being to save the remainder of the cargo, must have been successful and made by order of the Master.

Damage Control Parties

The activity of any damage control party will be decided by the nature of the emergency incident. The ship's Chief Officer would expect to deploy manpower where it will do the most good. Some damage control parties are designated beforehand – e.g. vent closure party in the event of an internal fire. However, the nature of the incident dictates what must be done and very often the order in which actions should be prioritised.

After a collision, men could well find themselves building a collision patch while others may be operating pumps to combat flooding. Actions would clearly depend on the severity of damage and the type of incident. If abandonment becomes a possibility then men could easily be deployed alternatively to turning out survival craft and getting them ready to launch.

Communications, both internal and external, will become an essential element of any incident. The Communications Officer goes to an immediate stand-by situation, ready to transmit urgency or distress messages as required by the Master or Officer in Charge.

Contact between the coordinating Chief Officer and damage control parties is maintained through walkie-talkie radios. The navigation bridge would, by necessity, be kept in the communication loop, as respective decisions must be made by the Master. The main engine control would also be linked via the 'conn' and the bridge position. Such arrangements would provide access to all the vessel's on board facilities.

Specialised equipment is often located at various stations around the vessel – e.g. in the UK, ro-pax vessels must now carry an emergency equipment locker above decks, on either side. (Ref. S.I. 1988 No 2272). It must be anticipated that the location of tools, ropes and similar equipment would be dispersed about the vessel for use by damage control teams. An example is the heli-deck landing area, which must carry a crash equipment box. Fire axes are also usually readily available at designated fire stations. The bridge will usually contain the emergency safety lamps, alongside portable radios.

NB. Ro-pax equipment lockers are expected to contain:

1 crowbar; 1 lightweight collapsible ladder (minimum 3 metres); 4 sets of waterproof clothing; sealed thermal blanket; 5 padded lifting strops for adults; 2 padded lifting strops for children; 3 hand-powered lifting arrangements; first aid kit; torches or lamps; 1 × 7 lb maul; 1 short-handled fire axe; 1 long-handled fire axe; 1 × 10 m rope ladder.

Figure 1.5 Example of emergency equipment locker.

Such lockers are subject to inspection by the examining surveyor prior to the issue of a Passenger Ship Safety Certificate.

Passenger Ship Collision

A collision between two vessels is a dramatic event in its own right, with its own associated problems. A collision at sea involving a passenger vessel raises the potential for greater loss of life and the ramifications are far more extensive.

The *MS Estonia* encountered heavy weather in the Baltic Sea in 1994, wind force 7–8 and wave heights of 4–6 metres. The bow visor to the ro-pax vessel separated from the ship and the bow door gave way. Subsequent flooding took place and the vessel listed initially by 30°–40°, then after about an hour she angled to 90°. A loss of life of 852 persons, mostly Swedish and Estonian citizens, occurred in this disaster.

Although this was not a ship-to-ship collision, but one of structural failure, the effects would be parallel to a passenger vessel sinking. In the case of the Estonia, the list would not have helped survivors evacuate the doomed vessel. But the weather conditions and the sea temperature in the Baltic, in September, would have been detrimental to survival.

Collisions are not planned and can occur at any time, anywhere and in any weather conditions. With this in mind, ship management and crew training must be effective.

In the event of a major incident, passenger vessels are vulnerable to incurring high loss of life because of the business they are in. Fortunately, major incidents are few, but the fact that they still occur is a reality, and the industry must be prepared for that one-off situation.

Figure 1.6 Passengers seen wearing lifejackets in 'crocodile formation'.

The Larger Passenger Vessel

Figure 1.7 The *Star Princess*, seen in Alaskan waters, in front of a glacier. The high freeboard and the position of the boat deck are prominent features of such vessels. Such reasons present the need for regular drills to prepare crew members for abandonment at any time. The ice scenario is an attractive but deadly adversary in the event of misadventure.

Managing an abandonment with large numbers of personnel will never be easy. Such an operation would need to be 'stage managed' and to do that crew members need to be well-practised in seamanship and crowd control. (Further information on the subject of abandonment can be found in Chapter 5.)

Tanker Collision

The tanker sector of the shipping industry is one of, if not the, largest of all the sectors. It is therefore feasible that a tanker loaded or in ballast has every possibility of being involved in a collision. History has provided us with relevant examples.

On 11 May 1972, off Montevideo, the *Roystan Grange* was in a collision with the Liberian tanker *Tien Chee*. Every member of a 61-person crew, plus 12 passengers, died on the *Roystan Grange*. Six out of a 40-man Chinese crew on board the tanker also died.

The collision occurred in thick fog conditions and the enquiry determined that once the tanker caught fire there followed several explosions and the flames were blown down towards the *Roystan Grange*. She was a refrigeration vessel using Freon gas as a refrigerant. The heat and ensuing fire probably burnt up all the oxygen, and persons probably died in their sleep from carbon monoxide poisoning. The *Roystan Grange* was towed to Barcelona, Spain, where she was scrapped in 1979.

Accidents occur when least expected and it may or may not be blamed on any one individual. Fog is frequently a contributing factor, as in the example of the *Roystan Grange*. The fact that it was a reefer vessel may have been a contributory factor also, but the involvement of the tanker and subsequent fire must always be considered as an inherent danger with such vessels.

In the event of a collision in which a tanker is involved, keeping the two ships close together without tearing damaged metal structures apart could possibly prevent ignition of the tanker cargoes' vapours. It is suggested that a blanket of foam could be beneficial to cover the contact area prior to deliberately trying to separate the two vessels.

Sea conditions may not allow the vessels to remain close together in any event, and a Master of a non-tanker would not want to remain in proximity of toxics or additional explosion risks to his own vessel. There are legal reasons for them to stay close together for mutual assistance, but a Master could not be expected to leave his vessel in close proximity to a known and expected danger.

Where fire is an outcome of the collision, the smoke from burning oil can be expected to be highly toxic. Dense black smoke, where present, can and will reduce visibility and fire fighters with breathing apparatus will experience great difficulty on approach to an oil blaze. Where power remains with the vessel, allowing it to be manoeuvred may allow the ship to be turned downwind to let smoke blow overboard rather than inboard. Such an action may prove beneficial to firefighters to attain an attack position in a reasonable smokeless area.

The release of hydrocarbons in the vicinity of personnel from whatever means is never a good thing. In the case of a tanker collision a release of oil or gas is uncontrolled and may have the side-effects of fire and toxicity for both crews to deal with.

Collision: Typical Damage/Repair Assessment (Hypothetical)

Overview

Following a collision in the English Channel in June 1996 the general cargo *M.V. Unlucky* suffered considerable impact damage on the starboard side, aft of the collision bulkhead and inward to the centre line of the vessel.

The main damage was incurred in the way of number one cargo hold to the outer hull plating between frames 75 to 115.

As this was a side impact, most of the damage was sustained by the outer shell plating.

Upper deck stringers were input and distorted to the edge of the hatch coaming, but the coaming itself suffered no structural damage. This would be confirmed by a monitored hose test when all other repairs are complete.

The keel was not damaged in any way, but the garboard strake, together with bottom plates and side strakes up to the sheer strake and the gunwale, have all experienced varying degrees of damage from the collision.

The following hull plates are required to be cropped and replaced:

- The garboard strake 'A' (11) in way of frames 95 to 100.
- Strake 'E' (16 to 19), crop residual and damaged plates and renew in way of frames 85 to 105.

The following plates need to be cropped, faired and renewed:

- Strake 'B' (8 and 9) in way of frames 90 to 100.
- Strake 'C' (6) in way of frames 85 to 95.
- Strake 'F' in way of frames 80 to 85 and 100 to 105; renew and joint all plates between frames 85 to 105.

- Strake 'G' in way of frames 80 to 85 and 100 to 110; renew and joint all plates between frames 80 to 110.
- Strake 'H' in way of 80 to 85 and 105 to 110; renew and joint all plates between frames 80 to 110.
- Strake 'J' in way of frames 75 to 80 and 105 to 110; renew and joint all plates between frames 75 to 110.
- Sheer strake in way of frames 75 to 85 and 110 to 115; renew sheer strake and gunwale between frames 75 to 115.

Miscellaneous

- Crop and replace fore peak pipe section in way of frames 95 to 105.
- Crop and fair deck stringers in way of frames 75 to 115.
- Crop and fair tween deck stringers in way of frames 80 to 110.
- Crop and renew bilge keel from frames 80 to 95.
- Crop, fair and renew sounding pipe in way of frame 85.
- Crop and replace deck gooseneck ventilator in way of frame 95, with associated pipe.
- Crop and replace the gunwale in way of frames 75 to 115.
- Scaffolding erected as required to effect all side repairs.
- Block removal to suit bottom plate repairs.

Additional

- Following inspection, plate keel remains intact.
- Replace starboard side, No.1, double-bottom tank plug.
- Watertight test to be placed on collision bulkhead, in way of frame 115.
- Pressure test double-bottom tanks in way of frames 60 forward to 115.
- Hose test deck stringer and hatch coaming connection in way of No. 1 cargo hatch.

Surface Coatings

- All renewed surfaces to be primed at paint shop.
- All welds to be re-coated.
- Respective protective coats to be applied as programme permits.
- Prime and coat with anti-foul coatings of all new steelwork between keel plate through strakes from 'A' to 'E' in way of frames 75 to 110.
- Coatings (boot topping) strakes 'F' and 'G' in way of frames 80 to 110.
- Coatings (topsides) strakes 'G' to sheer, in way of frames 75 to 115.

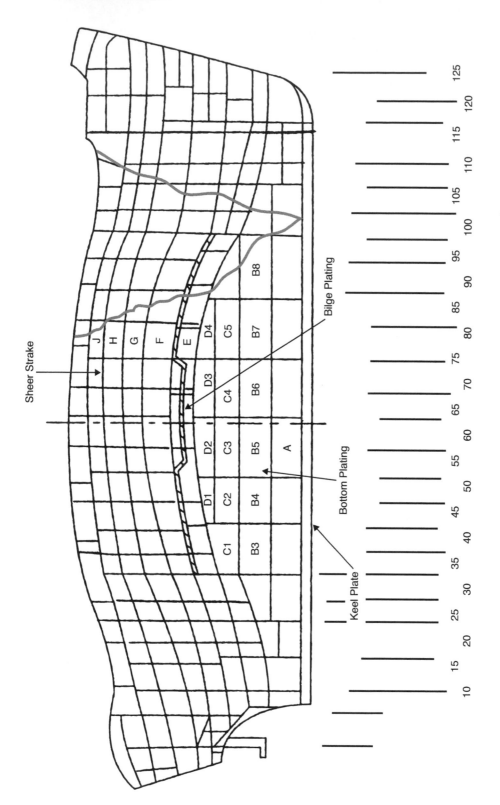

Figure 1.8 Shell expansion plan example (line diagram).

Summary

Following any incident where a vessel receives damage it would be expected practice that the company/owners would conduct a damage assessment by employing their own superintendents. This would be in virtually all cases, especially where the seaworthiness of the vessel is impaired. This would be further supported by a damage survey conducted by the Classification Society Surveyor.

The damage survey would detail the necessary repairs and these would be approved as they were completed by the repair yard inspection authority and finally by the Society's Surveyor. An interim Certificate of Class would then be issued to ensure that the damaged area has been repaired to the surveyor's satisfaction.

Full Certification of Class could be expected to be restored at the next complete scheduled survey of the vessel.

2

Taking the Ground

Grounding, Beaching, Stranding and Docking

Introduction

For a vessel to take the ground accidentally or intentionally is clearly not the ideal for a unit which is fundamentally designed to remain afloat. Most incidents of taking the ground could expect to lead to bottom damage to the hull, depending on the nature and the quality of the sea bottom.

In an incident where a ship runs aground accidentally, ships personnel would have to be concerned with the general arrangement of the vessel. The condition of the machinery space, for instance, is usually in a low-lying position and leaves it exposed to flooding. Any water ingress to the engine room could be expected to affect the ship's power supply very quickly thereafter.

Similarly, the ship's double-bottom tanks, often employed for fuel storage, are also generally situated low down, on either side of the keel. Seagoing personnel become familiar with the ships they sail on. Consequently, most would be aware that with fuel tanks lying low in the ship's hull the risk of pollution is real in the event that these double-bottom tanks are breached during a grounding incident.

Seafarers generally do not choose to take the ground and would never desire to take a rock-infested ground. Soft sand would be a better alternative in order to limit damage, in the event of being given a choice. Such a choice may present itself where a ship is deliberately forced into a shallow area, a process known as 'beaching'. This is a deliberate act usually carried out to prevent a total constructive loss of the vessel. Such a circumstance could occur, say, following a collision in which a ship is so badly damaged that she could sink. Beaching the vessel deliberately could save the ship from going down and becoming a total loss.

Whether running aground accidentally or deliberately beaching a ship, the incident is still an act of taking the ground, a totally alien situation for a vessel to find herself in.

Such a scenario is one that mariners may have to respond to at any time of the day or night if involved in an incident. What actions would the ship's Master have to adopt if the ship takes the ground? What actions would be expected of the Officer of the Watch and what would be expected of Chief Officers, Chief Engineers and other members of the crew?

Running Aground

Running aground is defined by a ship which strikes the sea bed and has no under-keel clearance. Running aground is accidental, whereas 'beaching' occurs where a ship takes the ground deliberately. A vessel can have a soft landing or a bad landing in a grounding scenario. These are usually separated by the amount of damage incurred to the vessel as it runs aground and will be clearly influenced by the nature of the sea bottom.

In any event, ships don't run aground in deep water. They run aground in shallows when their draught is too great for the available depth of water. The majority of groundings take place through poor navigation in restricted waters. Alternatively, if a ship experiences a power loss and subsequent black-out, the prevailing weather can cause the vessel to be blown aground (as in the case of the *Riverdance*).

There have been several high-profile disasters due to grounding incidents. The Finnish cruise ship *Sally Albatross* ran aground in the Baltic Sea in ice-covered conditions (March 1994); 1,250 passengers and crew were landed on the ice by a marine evacuation system (MES). This was the first time an MES had been deployed in anger. The grounding was put down to poor monitoring of the ship's position – in other words, questionable navigation practice.

The *Ulster Sportsman*, a ro-ro ferry en route from Ardrossan, Scotland, to Loch Larne in Northern Ireland ran aground at 16 knots, into the cliff face of the Irish coast. Although the keel was set back and concertinaed for about 1.5 metres, the main damage was to the bow. Fortunately, the collision bulkhead held the watertight integrity and prevented consecutive flooding. The ship, with tug assistance, was able to move to Belfast Dry Dock to effect repairs.

It should be realised that running aground generally occurs more frequently than collision. That is not to say a grounding incident is any less critical than a collision. The circumstances of a collision and respective damage incurred may be of a different nature and affect a different region of the ship compared to running aground. The seriousness and risk to life of both incidents could be similar.

In a situation in which the vessel runs aground, the duties of the Master and senior officers are more defined. As with a collision, the Master would move to the bridge and take the conn. He would also order the chief officer to obtain an initial damage assessment containing exactly the same four elements of a collision damage assessment, namely:

1 watertight integrity of the hull;
2 engine room, wet or dry;
3 casualty report;
4 any pollution incurred.

Clearly, the emphasis on the state of the engine room and whether it is wet or dry will reflect the position of the machinery space as it is so low-lying in the ship's structure. The same applies to the pollution observations, bearing in mind that fuel tanks are generally located in the double-bottom tanks, close to the structural region of contact.

Figure 2.1 The *Maanav Star*, a cargo vessel seen well aground in soft sand on a Sussex beach on the south coast of England. The vessel dragged her anchor and was subsequently blown ashore on 11 September 2004. After several unsuccessful attempts to refloat her the stranded vessel was finally refloated on 24 September with help from a Dutch tug.

Figure 2.2 The *Maanav Star*, aground as seen from astern.

Figure 2.3 Side elevation of the *Maanav Star* seen around on a beach in Sussex England.

Incident Report: Loss of the *Riverdance*

In January 2008 the ro-ro vehicle ferry *Riverdance* sailed on a short passage from Warren Point in Northern Ireland towards Heysham in Lancashire, England. This was a coastal journey she undertook daily. In the voyage across the Irish Sea the ship encountered some heavy seas and the ship's Master reported that the vessel became disabled by an encounter with what he described as a 'freak wave', in a position 10 miles west-south-west of Fleetwood, an area known as 'Shell Flat'.

The vessel lost power and steerage and listed to about 60° in the Morecambe Bay area. The cargo was caused to shift and the list settled to about 35°. The prevailing westerly weather caused the vessel to set down towards the north-west coastline of the United Kingdom. The Master generated a distress message and the four passengers and non-essential crew were evacuated by helicopter. The remaining crew members were later evacuated and the ship was allowed to remain at the mercy of the weather.

The *Riverdance* was subsequently driven aground on the Cleveleys beach, just north of Blackpool. The grounding initially did not appear to cause much additional damage as the vessel had landed on sandy, rock-free ground, with a gentle slope. First inspections led to an assumption that refloating the vessel was a distinct possibility. Although the region was tidal with 10 metre tides, it was still felt that the vessel could be refloated and saved.

The prevailing weather during February and March of that year remained unfavourable and the vessel, over a period of time, sank deeper into the sand.

All future hope of refloating the vessel seemed to be dropping away and a decision was made to salvage what cargo could be saved between tides.

A later decision, to accept the loss, was then made and the fuel oils within the vessel were recovered. Salvers were moved in to break the hull down by cutting it into moveable sections. This process lasted until September 2008 a period of over eight months.

> **NB.** The sequence of illustrations in Figures 2.4–2.6 show the vessel on first landing. The daily tidal effects and ferocious bad weather restricted salvage activity. The depth that the bridge wing submerged into the sand and the final resting place are depicted. From the final rest position, bunkers, cargo and engines were removed, prior to cutting the hull into manageable, transportable sections.

Initially the vessel was expected to be refloated and salved with its cargo at an appropriate high tide. The schedule was subsequently hampered by inclement weather generating heavy seas and causing the vessel to heel over considerably onto her beam ends. Further attempts to recover the ship were considered, but again bad weather worked against the salvage teams and the ship settled deeper into the sand.

The *Riverdance* was finally declared a total constructive loss and salvage experts were left with no alternative but to cut the ship into moveable sections for scrap value. The salvage operation was conducted by the removal of some of the accessible deck cargo being removed shortly after the vessel foundered. However, the bulk of the cargo was removed by cutting holes into the ship's side and craning cargo units clear of the hull prior to section cutting through the vessel.

Figure 2.4 The *Riverdance* ro-ro ferry buried in the sand north of Blackpool, England, prior to being cut into sections by salvage teams. Some idea of the depth that the hull is buried can be determined by the height of the exposed bridge wing seen above the surface.

Figure 2.5 The *Riverdance* seen aground and heeled to starboard surrounded by the incoming tide.

Figure 2.6 The *Riverdance* seen where the tidal water has receded. Some drop trailers are still seen under the accommodation structure on the upper vehicle deck.

Figure 2.7 The port side of the *Riverdance* seen in a setting position.

Figure 2.8 The *Riverdance* seen heeled over on her beam.

Figure 2.9 The *Riverdance* seen aground and with the starboard bridge wing settling deeper into the sand.

Incident Report: *MSC Napoli* (Container Vessel)

The *Napoli* experienced a hull failure after its departure from Antwerp on 17 January 2007. The vessel was outward bound on a course of 240° passing through the English Channel when she encountered large waves during the following day. A loud crashing or cracking sound was heard by the ship's personnel and the engine room alarms activated. Subsequent investigation revealed that a hull crack had appeared in the forward engine room bulkhead. This fracture was later confirmed by a diving inspection.

With the vessel's engine room flooding the Master, after consultation with the Chief Engineer and sighting cracks to the hull on either side, decided to abandon the ship. By 1125 hours the main electrical power had failed and the emergency generator had cut in automatically. The port lifeboat was prepared for launch because of the list on the vessel. Extra water, the SART and EPIRB were placed into the boat and the Master and Third Officer were the last to board the survival craft.

All 26 survivors were subsequently recovered by highline operations to two Sea King helicopters. After the rescue of the crew, the vessel was taken in tow towards Portland, Dorset. Concern increased regarding the condition of the disabled vessel as the operation approached the south coast of England. The possibility of the vessel breaking up or sinking became a real concern and it was decided to beach the ship in Branscombe Bay on 20 January 2007.

Some of the containers were lost on beaching the ship, but during the following five months the remaining containers were removed as was most of the ship's fuel oil. The vessel was refloated on 9 July, but it became readily apparent that she was in a poor condition and was quickly re-beached three days later.

After considerable deliberation the decision was made to break the vessel in two and salvage each half. This was to be achieved by explosive charges set each side and forward of the bridge front in way of the original cracks in the hull. The two halves of the vessel where then taken to be scrapped in April 2008, the stern section to Holland and the bow section to Harland & Wolf shipyard in Belfast.

NB. The *MSC Napoli* was built by Samsung Heavy Industries of South Korea in 1991. Her beam was too large to allow passage through the Panama Canal. At the time of building she was one of the largest container vessels to be built, accommodating 2,318 containers, 700 of which were stowed on deck at the time of the incident.

Figure 2.11 Bow section of the *MSC Napoli* in Harland & Wolf dry dock.

Figure 2.10 The forward section of the *MSC Napoli*, seen in the Harland & Wolf dock.

Figure 2.12 Hull section of the *MSC Napoli* at the position of breaking the vessel in two.

Beaching

The act of deliberately running a vessel into shallow waters to take the ground has become known as 'beaching'. The operation is usually only carried out to save the vessel from an even greater disaster like sinking and becoming a total constructive loss.

Ideal conditions for beaching a ship are usually very difficult to find all in one place and at the same time. Any Master finding himself/herself in such a predicament would expect to limit any additional damage to the hull by taking the ground. Hence, a soft, sandy beach must be considered the ideal position. The ground should preferably be rock-free, with a surf-free shore line.

Using limited speed, with maximum ballast on board, contact with the ground should be in daylight rather than during the hours of darkness. Beaching head first is expected because of the position of the collision bulkhead. Stern-first beaching would never be a recommended scenario because of the additional damage to rudders, propellers and hull that could be expected.

In terms of timing of a beaching operation, in relation to tides, it should not be on the top of high water of a spring tide. Beaching, if necessary, should occur at a mid-point between high water and low water on a falling tide. The vessel should avoid beaching on a rising tide because there remains the possibility that the vessel could be pushed further up the beach as the tide rises.

It should be borne in mind that beaching is a deliberate act and carried out with the view to refloating at a later date following repairs. If the vessel is pushed further inland by a rising tide, refloating may pose greater difficulties. It is recommended that

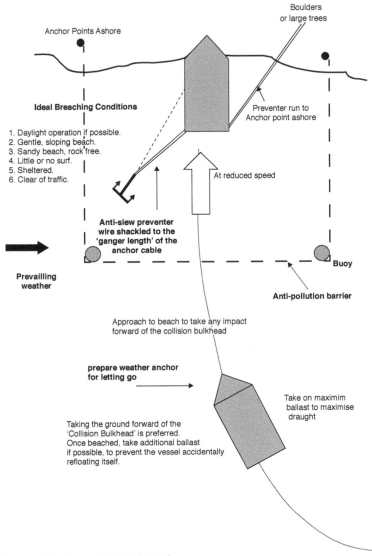

Figure 2.13 An approach to beaching.

once the keel has touched the ground, every effort to add further ballast weight to the ship should be made. This additional weight should limit unwanted movement of the hull to a position further inland. Additionally, both anchors should be walked back to stop the vessel accidentally refloating itself on a rising higher tide.

Once in a beached position in a tidal region, steps must be taken to prevent unwanted movement of the vessel by either tides or weather. Where moorings can be stretched and secured ashore these would act as anti-slewing, anti-movement retainers. Any future movement on the hull could only generate additional and unwanted damage to the vessel.

Improvisation remains one of the basic elements that stimulates so-called 'good seamanship'. Moorings and steadying lines would need anchor points to secure to. These could be provided by stalwart trees, large embedded boulders or even man-made anchor points.

Grounding/Beaching Summary

Carry out a damage assessment following the action of the ship taking the ground. The damage assessment should initially cover:

- watertight integrity of the hull;
- engine room check, wet or dry;
- casualty report for injuries;
- pollution assessment.

Subsequent actions are:

- sound round all internal ship's tanks;
- take a full external soundings with particular attention to the nature of the bottom and the forward and after end regions;
- deploy any remaining anchors;
- display aground signals as appropriate;
- seal the uppermost continuous deck (this should be achieved before taking the ground);
- maintain a deck patrol for fire and security;
- calculate the next high water/low water times and heights;
- investigate stability and refloating details following the instigation of repairs;
- prior to attempting to refloat, call in a 'stand-by vessel';
- ensure log book accounts are entered of all events.

Master's advice (once on the ground):

- order a position to be placed on the chart;
- following damage assessment results, open up communications to relevant authorities, including the Coastguard;

- engage tug assistance if appropriate;
- investigate damage and stability criteria as soon as practical;
- make a report to the Marine Accident Investigation Branch (MAIB);
- investigate the nearest dry dock capacity/availability/facilities with owner's assistance and/or instigate diver inspection;
- obtain a long-range weather forecast;
- monitor any ongoing repairs.

NB. Increased damage can be expected with any movement on the hull once the vessel has taken the ground. Any preventative measures, like the use of anti-slew wires and the deployment of anchors to reduce movement, are to be recommended.

Any change in direction of the prevailing weather as well as changes in tide heights can be expected to influence ship movement. Unless the ship is secured in position, a rising tide could cause the vessel to accidentally float off into deep water prior to instigated repairs being finished.

NB. It is imperative that an anti-pollution barrier is established around a damaged vessel. This can be achieved by obtaining designated boom equipment. If this is not available, then the use of floating mooring ropes could be employed as an improvised barrier.

Case Study: Running Aground

Incident

While on a voyage from Europe to Montreal during the winter of 1973, the container ship *M.V. Cast Beaver* (previously known as *Inishowen Head*) ran aground in fog just South of Quebec, in the St Lawrence river. Following a damage assessment, it was found that the underwater area of the hull of the vessel was seriously damaged, from stem to stern, after the ship had settled on a rocky river bed.

Although the watertight integrity, in way of the double bottoms, had been broached throughout the ship's length, the collision bulkhead and the tank tops had remained intact and the ship's positive stability had been retained.

Shipboard Actions

Immediately following the grounding incident the ship's Chief Officer was ordered by the Master to carry out a damage assessment. His activity was somewhat hampered by snow-covered decks and an extremely low outside air temperature of –20 °C.

Damage Assessment Content

Tank soundings revealed that all double-bottom tanks from No. 1s to No. 5s had been broached, but all cargo holds were dry. Visual inspection confirmed that tank tops were dry and cargo was undamaged. The engine room was dry but double bottoms containing fuel oil had been broached. This was also evident on the upper deck as fuel oil was being forced up air and sounding pipes by incoming water pressure. This oil content on the upper deck was visible from the navigation bridge. There were no injuries to crew and no signs of pollution overside.

Immediate Actions

The Master took the 'conn' of the vessel and made an immediate urgency call for tugs to assist either side of the vessel. He ordered a position to be placed on the navigation chart and obtained an updated weather forecast. The Chief Officer ordered the deck scuppers to be blocked off and ordered all crew members to observe a non-smoking ban. He then carried out a detailed stability assessment and reported to the Master that the vessel would remain afloat on the tank tops at a reduced but positive stability.

Subsequent Actions

Following urgent discussions with shore-side authorities, the availability of the Quebec dry dock was confirmed. This dock was seen as a suitable alternative to carrying out a detailed underwater inspection.

The ship was de-ballasted and an interim passage plan was established to move the damaged vessel, floating on tank tops, with tug assistance, towards the dry dock entrance. The shipping company and agents were informed and the vessel entered dry dock on 5 January 1974. The company superintendant made immediate plans to attend the ship in the Quebec dry dock.

The Dry Dock Operation

The ship supplied the dock with relevant documentation, i.e. dry dock plan, general arrangement and shell expansion plan. The

NB. The Dry Dock Authority advised the ship's Master that because of the extreme cold weather, inherent dangers could affect the wellbeing of the propeller area of the vessel once exposed out of the water. It was further explained that the propeller area was constructed with dissimilar metals and such cold temperatures could generate brittle fractures in extreme conditions.

block pattern was set in place prior to the vessel entering the dock in the late evening.

Once the ship had become sewn on the blocks a scaffold enclosure was erected around the aft end. This was given a polythene surround and industrial heaters immediately brought in to keep an ambient temperature in the areas of the propeller and the tail shaft. These precautions proved necessary to prevent metal fractures in temperatures that were to drop below $-30\,°C$.

The First Morning

General inspection of the lower part of the hull revealed that shell plating from the garboard strake, either side of the keel, had been badly torn through the vessel's length. Immediate activity started to crop away torn metal sections and fair others. The ship's Second Officer was designated to take an account of all the new steelwork used to renew the damaged hull. This was a company requirement to determine the amount of new steelwork that would be needed to repair the damaged hull.

Aftermath

The hull was repaired in just over four weeks, after which the vessel was restored and refloated. She was able to continue onto Montreal, discharge and load her cargo and returned to her home port of Liverpool by the second week of March 1974. The estimated cost of this major repair was given as $2,000,000.

Soundings and Use of Lead Line

In virtually every case of grounding, two sets of soundings must be taken. The first is a full set of internal tank soundings, the other is external depth soundings. Some mariners might feel that the use of the hand lead is dated and would generally never be employed because in this day and age we have the benefit of excellent echo sounding, sonar and Doppler technology. In routine sailing operations such opinions would be right, but 'smelling the bottom' is not generally considered a routine operation for ships.

Where navigation historically led ships into shoal waters the hand lead was always used, the lead being swung from a position known as the chains. This operation of 'swinging the lead' was conducted when a vessel entered estuaries and engaged in river passages. It is in such positions, where shallows could be expected, that the vessel's draught could be compromised.

A typical geographic example today could be found in and around the hinterland of Central American countries of El Salvador, Nicaragua and Costa Rica. Any vessel moving into estuaries' could expect to be engaged in continuous position monitoring and continuous depth readings, with possibly less than 1 metre depth under the keel. The monitoring of under-keel clearance in such circumstances is considered essential for navigation in shallow waters.

However, once the vessel takes the ground, the possibility of damage to sounding equipment is a reality. If already aground, with zero depth, depth gauging instruments will no longer be of any practical use. The need for the 'hand lead' comes to the fore. The function is twofold: one to provide the depth of water that the ship lies in; the other to determine the nature of the sea bottom that the vessel lies on.

Table 2.1

Marks on Fathom Line	Fathoms	Depth	Metres	Marks on Metric Line
			1	1 strip of leather
Piece of leather with 2 tails	2		2	2 strips of leather
Piece of leather with 3 tails	3		3	Blue bunting
			4	Green & White bunting
Piece of white linen	5		5	White bunting
			6	Green bunting
Piece of red bunting	7		7	Red bunting
			8	Yellow bunting
			9	Red & White bunting
Piece of leather with a hole	10		10	Leather with a hole in it
			11	1 strip of leather
			12	2 strips of leather
Piece of blue surge	13		13	Blue bunting
			14	Green & White bunting
Piece of white linen	15		15	White bunting
			16	Green bunting
Piece of red bunting	17		17	Red bunting
			18	Yellow bunting
			19	Red & White bunting
Piece of cord with 2 knots	20		20	Leather with hole + 2 strips

Taking soundings around the ship to determine depth and nature of the bottom is customary, in particular around areas of the bow and stern. In a grounding situation it is often the case that the bow of the vessel will become the first area of contact and the ship's Master would want to know the nature of the ground he is on. Soft sand is obviously better than a rocky sea bed. The nature of the sea bottom is obtained by arming the lead (tallow or soft soap pressed into the arming recess at the base of the lead). The additional need to ascertain depth at the stern is to determine whether the propellers are in clear deep water and can be turned. The need to use propellers is influential in any decision to attempt to refloat.

On occasion a vessel may take the ground when alongside a berth and loading cargo. Where the tide is falling, vessels have been known to sit on the bottom at low water. Soundings taken on either side of the vessel will give an indication as to the ground level and whether the vessel is on the slope of a shelf or a flat region of the sea bed.

In all such cases, the use of the echo sounder is restricted and the lead line becomes the old-fashioned instrument of the day. The only change that has occurred since the days of sail is that lines are now employed with metric marks, but many ships still carry the historical lines calibrated in fathoms. Ships are still legally obliged to carry a hand lead, which could be measured in fathoms or metres.

A lead line is 25 fathoms in length, constructed in cable-laid hemp. The lead itself weighs 7–9 lbs and is about 9 inches (23 cm) in length and fitted with a becket strop to attach the eye splice of the line.

The line is measured from the eye splice and the additional length from the splice to the base of the lead is known as the 'benefit of the lead', being to the benefit of the ship.

Emergency Dry Docking

After any incident which has incurred shipboard damage, especially hull damage, through grounding or collision the vessel will invariably be classed as being unseaworthy. Such a state would render insurance null and void and the ship would effectively lose her class.

The owners and the Master would look to effect repairs as soon as is practicable and apply to obtain an 'Interim Certificate of Class'. Where high levels of critical damage have occurred the services of a dry dock are usually required, even if only for a 'bottom' inspection. The severity of the incident will dictate the nature of inspection and subsequent repair activities. By comparison to a major broach in the hull, if the vessel touched bottom in soft mud or sand, this would probably warrant an underwater inspection by divers. This would be especially so where internal tank soundings have remained unchanged, to ensure peace of mind.

Dry docking operations for the most part are considered routine to ensure ships are retained and operated at safe, legislated standards. The docking periods provide surveyors access to all parts of the vessel, including the underwater volume.

When the routine docking schedule is interrupted by an emergency incident, most shipowners would expect to amalgamate emergency repairs with any routine survey work to avoid docking again too soon. Any docking period is one that costs the ship owner not only the monetary expense but also the loss of income from loss of trade.

The principle of dry docking a ship has the same stability effects as when a vessel runs aground. The vessel taking the blocks in dry dock will experience an upthrust through the keel which will effectively cause the centre of gravity (G) of the vessel to move upwards, towards the metacentre (M). This force, known as the 'P' force could cause the vessel to become unstable during the docking period if G is allowed to move above M, generating a negative stability condition. Such a situation could cause a capsize situation when taking or leaving the blocks, with serious consequences for the ship and personnel on board.

Where a vessel runs aground, that same P force is experienced through the keel position and the vessel loses her buoyant state. The positive stability of a ship going into dry dock is the responsibility of the ship. The ship's Chief Officer is tasked to ensure that positive stability is retained throughout the critical period and he would usually increase the range of the GM to compensate for the effects of P force.

Docking Variations

The Graving Dock

This is a large ship dock fitted with a caisson access from an active deep water area. They are usually constructed in stone or concrete with stepped sides known as 'altars'. Built blocks fitted with softwood landings are laid in a pattern on the inclined dock floor. The block pattern corresponds to the individual ship's 'docking plan'.

The dock space is allowed to flood to a level equal to the external outer waterway level. The caisson (dock gate) is opened to allow the vessel passage to enter the docking space once levels either side of the caisson are equal. Once inside the dock, the caisson is closed and the vessel is held to a position over the block pattern. The water is then drained via the declivity of the dock, allowing the vessel to become sewn on the blocks.

Block Pattern and Shores of the Graving Dock

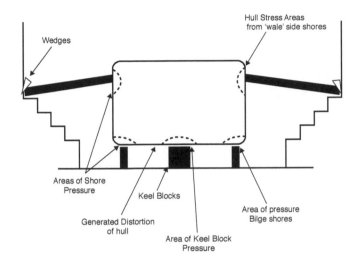

Wedges

Hull Stress Areas
from 'wale' side shores

Areas of Shore
Pressure

Keel Blocks

Generated Distortion
of hull

Area of Keel Block
Pressure

Area of pressure
Bilge shores

Figure 2.14 The outline of a ships hull seen supported in dry dock, by bottom shores and side (wale) shores. Stress areas indicated where shores make contact with the outer hull.

Where the beam of the vessel is large, intermediate blocks may be deployed between the bilge shores and the keel blocks. Wale shores might not be required. Where blocks and shores are positioned against the hull, a level of plate distortion is to be expected. This distortion is generally not a permanent concern as the shell plate will assume its natural form once the vessel works itself in a seaway once re-floated.

Floating Dock

Floating docks, because of a generally restrictive size, have a tendency to cater for the smaller tonnage and specialist-type vessels. This is not to say that some of the larger floating docks do not have the capability to offer docking facilities to larger tonnage vessels, but they are not usually occupied by the VLCC/ULCC class of ship. The small coaster, dredgers, fishing vessels, research vessels, etc. are the more usual customer for the floating docks.

The ability of the floating dock to be towed into alternative locations is a distinct advantage and one that has often been employed within the salvage field. They are also regularly employed in association with the larger shipyard facility to provide continuous working for secondary vessels, while at the same time keeping the larger docking feature readily available for the 'big order' option. Having said this, it should be remembered that they are continually immersed as a working unit and the floating dock itself often requires being 'dry docked' for essential maintenance to ensure continuous operations.

Figure 2.15 The passenger vessel *Minerva* seen in the floating dock at the Dubai docking complex. Many dry dock yards carry additional docking space to accommodate additional work when the main graving dock(s) is occupied.

Figure 2.16 Floating dock arrangements.

Floating Dock Structure

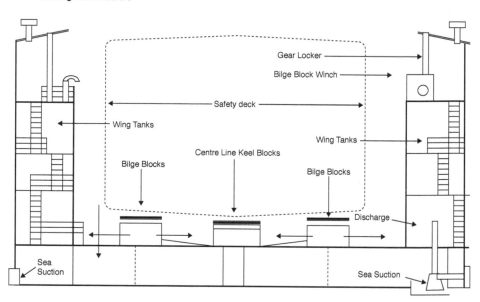

Gear Locker

Bilge Block Winch

Safety deck

Wing Tanks

Wing Tanks

Centre Line Keel Blocks

Bilge Blocks

Bilge Blocks

Discharge

Sea Suction

Sea Suction

Figure 2.17 A floating dock seen in a working operation, with a vessel exposed above the waterline. The two sides of the dock and the volume under the dock floor contain the tanking system, which can be flooded or pumped dry, allowing the passage of the vessel inward or outward from the docking space.

Synchro-Lift Docking Operations

The 'Synchro-Lift' method of docking must be considered as probably the most modern style of docking a ship within the marine industry today. It is certainly also the most flexible, in that it does not occupy the dock space and subsequently does not deny the use of the facilities to other vessels wishing to dock. These up-to-date facilities can subsequently dock and handle a greater number of vessels at any single time without having the operation compromised with a single-ship occupation of the dock.

They do have their limitations, however, in that they tend to have a maximum lifting capacity, whereas the larger graving docks can accommodate ships in excess of 500,000 dwt. Another concern would also be for loaded vessels, which would be affected by a lack of ability to place shores around the hull when in an open parked, docked scenario. Increased stresses from cargo load could clearly cause excessive hull distortion in such circumstances and would necessitate almost certain discharge of cargo from the larger vessel prior to docking being allowed.

Against the disadvantages are considerable advantages which go along with multiple dockings. A regularly employed workforce is always available and desirable, working on several ships over an indefinite period of time. Working constraints and time schedules are independent to individual ships, whereas when two or more ships dock in a single graving dock each depends on the needs of the other before flooding can take place.

The type of docking park that follows the lifting element tends not to have the constraints of the enclosed graving or floating docks which are in more common use. The hull is clearly more exposed and more accessible by all manner of vehicles/cranes etc., as well as presenting a lighter and cleaner environment for shipboard operations. This said, many ship operations present the same open aspect for exposed hull operations, but tend generally not to have the capacity or ability to accommodate the needs of the larger ship.

There is clearly more to go wrong and increased maintenance for a synchro-lift operation than for a floating or graving dock. The platform, the numerous lifting winches, together with the side pull winches, as well as the caisson, all require regular maintenance schedules. Lack of such schedules would clearly bring about a cessation of docking operations very quickly and the loss of confidence of potential customers.

Operation of Docking

The ship to be docked is hoisted on the rail bogey platform. The hoist marries with the extended rails at ground level to permit the ship on the platform to be rolled forward, clear of the dock. The

Figure 2.18 Exposed docking procedures by synchro-lift operation.

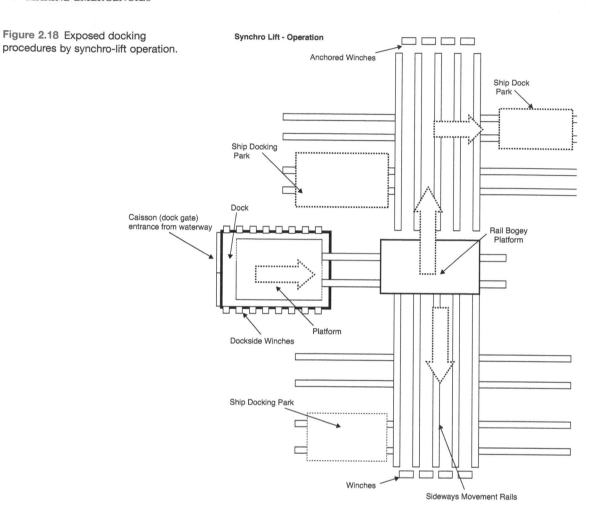

Synchro Lift - Operation

Anchored Winches

Ship Dock Park

Ship Docking Park

Caisson (dock gate) entrance from waterway

Dock

Rail Bogey Platform

Platform

Dockside Winches

Ship Docking Park

Winches

Sideways Movement Rails

Figure 2.19 The tanker *Stolt Victor* seen exposed at ground level in a ship bay at the Tandanor Shipyard in Buenos Aires, Argentina. This docking complex operates one of the largest synchro-lift operations in the world, having a lift capacity of up to 10,000 grt.

platform is then heaved sideways to place the vessel into a ship/docking park space. This system allows up to 12 ships to be docked at once.

Hydrolift Docking Systems

In December 2000, the Lisnave Shipyard at Setúbal, in Portugal, opened a platform hydrolift docking system for ships. This new concept allowed for the docking of three Panamax-sized ships simultaneously.

The system works in conjunction with a wet basin which is entered via a caisson at the seaward end. Once the inward bound ship is established in the basin, the outer caisson is closed. The water level is then increased in both the basin and the designated platform dock by pump operation. The platform dock can alternatively be filled by gravity from the basin area.

Once the basin is full and level with the dock space, the ship will have been elevated sufficiently to clear the sill of the platform dock. This position allows the dock caisson to open and permit the transfer of the vessel from the basin to the dock platform. The dock caisson is subsequently closed and draining of the dock is then allowed by gravity.

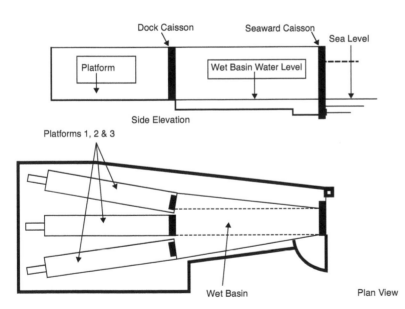

Figure 2.20 Side elevation and plan view of the hydro-lift docking complex at the Lisnave Shipyard.

Figure 2.21 The vessel *Guadalupe Victoria II* lies on the centre platform of the hydrolift docking system at Lisnave, Portugal as another vessel enters the secondary platform from the wet basin.

Figure 2.22 A block pattern laid to the platform in the hydro-lift docking system. The dock will be flooded and caisson opened to allow the expected vessel to enter the platform area.

Hull Forces: Grounding and Docking

Taking a ship into dry dock will generate an up-thrust force through the keel when she takes the blocks. This force is commonly known as the *P* force, which initially provides an apparent movement to the ship's centre of gravity (*G*), causing an effect of a reduced metacentric height (GM). The same force will be experienced by a ship which runs aground.

The up-thrust from the ground acting through the keel position is effectively that same *P* force that the ship would experience when docking. The danger in a docking situation is that *G* could experience an apparent move so high as to go above *M* (the metacentre) and generate an unstable condition. Such an apparent movement could cause the vessel to slip from the blocks.

Where a vessel takes the ground the area is usually of a greater, flatter surface than the small landing area of docking blocks. Such a landing would not normally have the inherent dangers of slipping to an angle of heel, as with slipping from the blocks.

The obvious ramifications of running aground will generally be associated bottom damage, whereas taking a ship to land on docking blocks, one would not anticipate causing any hull damage whatsoever. Grounding frequently involves loss of watertight integrity, which will directly influence the positive stability of the vessel. Any ingress of water would take the place of an added weight and also be coupled with the disadvantages of free surface movement, changing the GM value.

Calculating loss of GM

Two methods of ascertaining loss of GM and *P* force are:

$$\text{Force } P \text{ (tonnes)} = \frac{\text{MCTC} \times t}{L}$$

Where: MCTC represents the moment to change trim 1 cm; *t* represents the trim in centimetres on entering the dock; *L* represents the distance between the centre of flotation and the vertical line of action of the *P* force in metres.

The first method considers an effective rise in KG:

$$\text{Virtual loss of GM} = \frac{P \times KG}{W - P}$$

Where: *W* represents displacement of vessel; *KG* represents distance between the keel and the ship's centre of gravity.

The second method considers a reduction in KM:

$$\text{Virtual loss of } GM = \frac{P \times KM}{W}$$

Where: *KM* represents distance between keel and the metacentre.

3

The Lee Shore and the Use of Emergency Anchors

Introduction

Seagoing has always been a hazardous business both in good weather and in bad, but by far the majority of catastrophes occur during periods of bad weather. Situations are often made worse by bad weather being coupled with equipment failure or malfunction of main engines. In any event, to be caught off a lee shore and experience a power failure at the same time is not a good situation for anyone to find themselves in.

Options to keep the vessel safe are limited. The ship with no power tends to lose all navigation ability and may find there is a total loss of all communication systems. It must be considered one of those times that the emergency generator needs to be brought into play as it can become an essential element of survival.

Such a situation places the ship's navigation department at a complete loss and leaves the engineering personnel working overtime to rectify faults and effect repairs, as quickly as possible.

The threat of a rocky shoreline and the vessel facing a total constructive loss is a real one. Even if the vessel is not wrecked, a grounding experience would be the very least that could be anticipated. The role of the ship's Master becomes one of positive response to any incident, but without engines being restored, positive action may not be a possibility.

Various parameters can and will dictate what Masters can and cannot do. A day- or night-time experience will influence any activity. A mobile phone on board the vessel could easily become a saving grace when shipboard communications are down. Distance from shore and the weather state could directly affect the time interval of any effective response. So it becomes clear that each differing scenario will require a unique solution – one that is not always easy to establish in the face of adversity.

What is the Lee Shore?

In routine sailing the lee shore is a term given to a shoreline that is directly threatened by the prevailing weather. In normal circumstances any vessel would want to keep well out to sea and not expect to run the risk of being blown onto the shore. However, the term 'lee shore' has become synonymous with a vessel losing control of main engine power, or loss of steering through the malfunction of steering motors or loss of the rudder. In the event of a loss of engines or steering, the vessel could inevitably find herself at the mercy of the weather, being blown towards the lee shore.

Loss of Steering

The Officer of the Watch would probably be the first to experience any loss in steerage. This would be noticed by the ship's 'off course alarm' being automatically operated, assuming it is switched on in the first place. The ship's head would change alignment with either land or celestial bodies (e.g. the moon), moving quickly from one bow to the other.

It must be considered normal practice for the Officer of the Watch to report the loss of steering immediately to the Master. Such an incident would expect to result in the Master taking the 'conn' of the vessel and immediately bringing the vessel to a stop.

NB. This action could be similarly compared to driving a car on a motorway or freeway, and having the steering wheel come off. Would the driver increase speed or stop?

Once the vessel is stopped, an investigation as to the cause of steering failure would no doubt ensue. Where one steering motor is in operation, it would be practical to switch on the second to see the response. The ship's engineers would be engaged to initially fault find, then instigate repairs if possible.

Following investigation, assuming the vessel still has power, an inspection of the rudder itself could provide an answer. It is a matter of record that vessels have actually experienced the loss of a rudder. Such an occurrence would probably cause the duty engineers to experience a greater or lesser degree of 'cavitation' in the stern area of the hull. Excessive vibrations and/or screw race might be another symptom that all is not right in the region of rudder(s) and propellers.

Steering Gear Operations

Steering methods have changed considerably over the years, but the basic function has remained and that has been to move the rudder to change the movement of the ship's head. Even this basic function has had effective competition from the rotatable thrusters, controllable water jets and the more recent steerable pod technology. However, new concepts have not yet completely dominated the steerage of ships and many vessels are fitted with conventional rudder movement in order to control the vessel's heading.

Four Ram Electro-Hydraulic Steering System

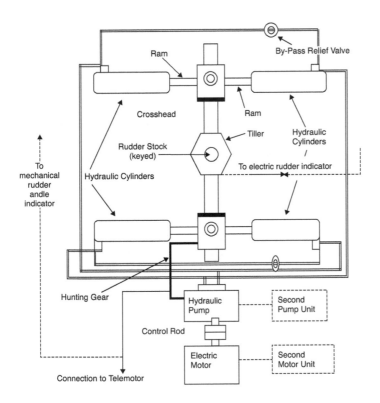

Figure 3.1 Rudder is keyed to the rudder stock of the tiller arm. Movement of the rudder is achieved by the four (4) rams acting on the tiller.

Compass Reliability

In association with loss of steering, a vessel can more frequently experience the illusion of loss of steering. Such an incident could happen if and when the gyro compass fails. Practically the automatic pilots are operational with a direct interface from the gyroscopic compass. Should the compass fail, the effect would be that the ship's head would probably be caused to wander, giving the impression of steering failure. The majority of seafarers would maintain that the compass is the most important of instruments on the ship. As such, a level of respect for a reliable gyro compass is a worthwhile consideration. The magnetic compass is, of course, always the fall-back fundamental of safe navigation.

Compass Failure: Ramifications

NB. Gyro compasses can of course be restarted, but settlement can take up to six hours before the compass would be considered reliable enough for use.

With the ship's gyroscopic compass interfaced into autopilots, radars, ECDIS and other plotting aids, the effects of a compass failure can be detrimental to the safe operation of the vessel. Rough weather and subsequent ship movement can cause the three degrees of freedom of the gyro to be adversely affected. Older models of gyro compass were known to topple, causing the immediate loss of autopilot and a reversion to the magnetic compass, with manual steering being applied.

Open-Loop Control System for Steering Operations

If the helmsman is considered as trying to steer a straight course, it will be realised that he compares the measured value (MV) of the ship's head with the desired value (DV) of the intended course. If these two values differ, then an error exists and the helmsman will apply corrective action by turning the ship's wheel (manual steering). The action of the helm is made opposite to the direction of the error.

This system does not ascertain the error, nor will it use the error to initiate corrective action. Once the helmsman is introduced, he/she carries out both tasks of ascertaining the error and applying the corrective action.

Figure 3.2 Functional elements of open loop control system of steering gear.

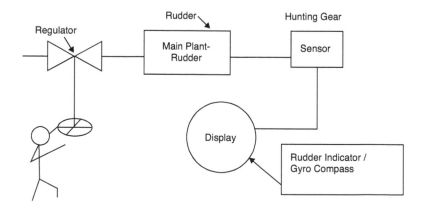

Steering Gear: Control of Transmission (Closed-Loop Control)

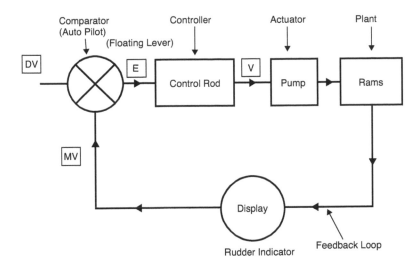

Figure 3.3 Functional elements of automatic (closed loop) steering operation.

With automatic, closed-loop control the comparison between the DV and the MV is made by a comparator within the system itself. The output from the comparator where DV and MV are not the same generates an error signal (E) which is passed to the controller. This amplifies the error signal and outputs a power signal (V), which can be used to apply corrective action by causing the pump and rams to be moved.

A feedback system which incorporates a rudder indicator on the bridge, displays the action and movement of the rudder. The hunting lever moves between the rudder stock and the control rod for the pump. This effectively switches off the power when the rams come to rest. Actual movement of the rams is transmitted by means of a control rod to a floating lever, the inner end of which acts initially as its fulcrum, which in turn is secured to the tiller by a rocker arm. Any movement of the control rod is transmitted from the opposite end of the floating link, through gearing to operate the valves of the hydraulic pump units, causing rudder movement.

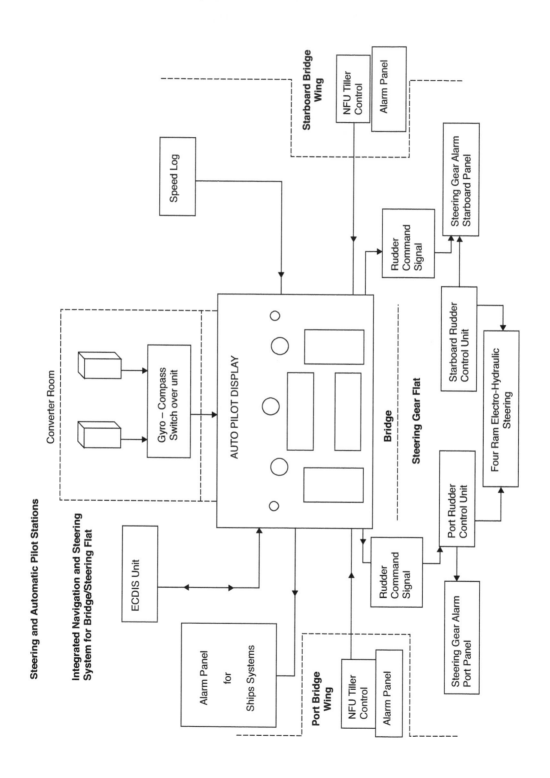

Figure 3.4 Navigation and steerage transmission between the ships bridge and the steering flat to generate rudder movement to meet navigational requirements.

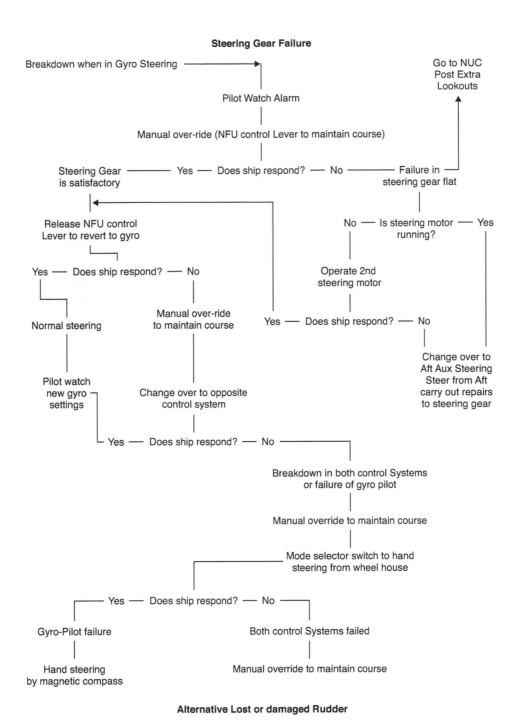

Steering Gear Failure

Breakdown when in Gyro Steering ⟶

Go to NUC
Post Extra
Lookouts

Pilot Watch Alarm

Manual over-ride (NFU control Lever to maintain course)

Steering Gear ⟶ Yes ⟶ Does ship respond? ⟶ No ⟶ Failure in
is satisfactory   steering gear flat

Release NFU control
Lever to revert to gyro

No ⟶ Is steering motor ⟶ Yes
          running?

Yes ⟶ Does ship respond? ⟶ No

Operate 2nd
steering motor

Normal steering

Manual over-ride
to maintain course

Yes ⟶ Does ship respond? ⟶ No

Pilot watch
new gyro
settings

Change over to opposite
control system

Change over to
Aft Aux Steering
Steer from Aft
carry out repairs
to steering gear

Yes ⟶ Does ship respond? ⟶ No

Breakdown in both control Systems
or failure of gyro pilot

Manual override to maintain course

Mode selector switch to hand
steering from wheel house

Yes ⟶ Does ship respond? ⟶ No

Gyro-Pilot failure

Both control Systems failed

Hand steering
by magnetic compass

Manual override to maintain course

Alternative Lost or damaged Rudder

Figure 3.5 Flow system for fault finding to steering systems.

Rotary Vane Steering Gear

Figure 3.6 The stator casting of the rotary vane steering unit, seen at the head of the rudder post in the steering flat.

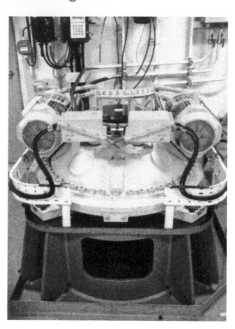

Figure 3.6 The stator casting of the rotary vane steering unit, seen at the head of the rudder post in the steering flat.

Rotary Vane Steering

Rotary Vane steering is a compact steering unit which is situated on top of the rudder stock. A rotor is 'keyed' onto the stock and the whole is encased by a steel casing known as a 'stator'. The concept allows follow-up and non-follow-up modes to operate with either electric or hydraulic transmission systems.

Model variations allow rudder angles of $2 \times 35°$ or $2 \times 60°$, with options of up to $90°$. The system tends to act as a self-lubricating rudder carrier, as well as generating the turning movement to the rudder. This is achieved by oil being delivered under pressure to one side of the blades of the rotor. With the rotor being 'keyed' to the rudder stock, when the rotor is caused to turn so does the stock.

Clearly the direction of turn will be effected by the direction of the pressurised oil affecting the rotor blades. Therefore, in theory the rotor and stock can turn only one way, namely in the direction of the pressurised oil. However, if the directional flow of the oil is reversed, by reversing the rotation of the oil pump, then the rotor will also be caused to turn in the opposite direction. This pump reversal from one direction to another provides the necessary directional oil flow to cause movement to port and starboard.

The oil under pressure is kept contained within the unit by the stator. The stator is dynamically sealed and is leak-free, generally providing an effective, alternative steering mechanism within the created pressure chambers.

Lee Shore: Loss of Main Engine Power

A dead ship at the mercy of the weather is a bad scenario; a dead ship at the mercy of the weather off a 'lee shore' is a potential major disaster. Such instances have led to a total constructive loss of the vessel coupled with a high possibility of loss of life.

When a ship blacks out and loses power, all control is immediately lost. Communications go down, steerage drops away and positive movement in any direction is eliminated. The only saving grace is the emergency generator, which will usually, on the majority of ships, automatically cut in. The emergency power source is designed to supply emergency lighting circuits and all essential navigation equipment, inclusive of: gyros, radars, steerage, navigation lights, etc. Additionally, emergency disembarkation lights cover the embarkation points of survival craft in the event that the situation deteriorates.

Loss of generator power can usually be dealt with reasonably quickly by use of a secondary auxiliary generator or from an independent emergency supply. A mechanical breakdown of the main engine, as in an engine room explosion, may prove a greater difficulty to handle, especially where spare parts are not available.

Clearly an inspection by engineers would need to take place as soon as possible, with an initial report designated to the possible repairs. Where it is found that damage is beyond the capability of the ship's engineers, the immediate problem returns to the bridge for alternative action to save the wellbeing of the vessel.

Master's Options

Loss of Control: No Steering

1 Master takes the conn.
2 Vessel displays signals for being 'Not Under Command' (NUC).
3 Order damage control assessment – check the second steering motor.
4 Place a position on the chart and enter it in the log book.
5 Establish a bridge team to include lookouts and helmsman on stand-by.
6 Clear away anchors for emergency use.
7 Assess tidal details for the region.
8 Prepare urgency/distress communication.
9 Obtain current weather report and expected weather report.
10 Monitor any ongoing repairs.
11 Investigate stern – bore, options (depends on proximity of land).
12 Investigate 'jury rudder' or 'jury sea anchor' options.

13 Make prudent use of ballast operations to ballast the fore end to establish flag effect on the ship's aft accommodation block (if the vessel can be re-oriented and engines are still available, positive headway into the wind may provide additional time before touching the shoreline).

14 Open up communications to bring a tug on stand-by.

Loss of Control: No Main Engines

1 Master takes the conn.
2 Vessel displays NUC signals.
3 Obtain written statement from the Chief Engineer on the engine status.
4 Place a position on the chart.
5 Establish a bridge team in situ.
6 Clear away anchors and walk back an anchor as depth decreases.
7 Assess tidal conditions for the region.
8 Ballast the fore end to provide flag effect to reduce windage on the ship's hull.
9 Open communications to establish tug assistance on stand-by.
10 Inform owners (not a legal requirement but a request for support).
11 Inform Marine Authority and file an incident report to MAIB.
12 Obtain an immediate weather report.

Additional subsequent actions may include:

• prepare Urgency message;
• turn out survival craft;
• stand-by windlass for possible deep water anchoring;
• prepare fore deck to receive and secure possible tug's towline;
• continue to monitor position and prepare distress signals;
• carry out detailed chart assessment of proximity shoreline;
• brief crew for possible abandonment.

NB. If anchors are to be gainfully employed, Masters should be aware that the capability of the windlass is to be able to lift 3.5 shackles of cable plus the weight of the anchor when hanging in the vertical. Such standards are for when the equipment is new and unworked on first installation. Older ships with well-used and worn anchor-handling facilities may not have the pristine capabilities of a new ship.

Relevant Anchor Work

Anchors come in all shapes and sizes, but each is usually functional to provide a level of holding power. Conventional shipping tends to operate with conventional stockless anchors (holding power being four times its own weight); large ships may be equipped with high holding-power anchors like the AC14 (holding power ten times its own weight). Changes in construction from cast steel to prefabricated steel provide the differences in holding ability.

Other anchors have dedicated operations as mooring anchors, reaction anchors or as piggy anchors within the offshore fraternity and the marine environment generally. Most commercial vessels operate with anchor chain stud cable, but many of the high speed craft and small ferry operators work with a wire anchor warp.

Virtually all non-military vessels work anchors by a windlass or heavy-duty winch arrangement. Some employ cable holders, which is a system common to some military craft, but in the main merchant ships conduct anchor work by a centre line or split port and starboard windlass operations.

Figure 3.7 Half windlass as fitted to the port side of a modern passenger vessel. A similar unit is fitted on the starboard side (not shown). The chain cable is seen with the 'bow stopper', compressor type, forward of the split windlass.

Anchor Use

Many marine incidents find the need to employ anchors, typically in the aftermath of running aground or holding off a 'lee shore'. Commercial vessels are equipped with two working bow anchors and these are a condition of a ship being classed as seaworthy. With this in mind the loss of an anchor compared with loss of the ship would seem a small price to pay – the vessel could always obtain another anchor for one that might be sacrificed for the wellbeing of the vessel.

Many vessels carry a spare anchor to ensure against a lost anchor and the possibility of losing her seaworthy class. Other ships, which don't carry a spare anchor, would probably not be allowed to sail unless issued with a dispensation from the Marine Authority. Such flexibility would normally only be allowed to reach a designated port where a replacement anchor could be supplied.

The offshore industry are frequently employed in laying and recovering anchors deployed in a pattern format for the purpose of securing rigs and installation positions.

NB. One of the most regular marine insurance claims is for lost anchors.

Figure 3.8 The open deck of an anchor-handling vessel seen recovering a conventional anchor.

Anchor Types

- Admiralty class (cast) anchor (AC 14)(AC 16A) (AC 17)
- Admiralty mooring anchor (e.g. type AM12 = 6 tonnes)
- Admiralty pattern anchor (fisherman's anchor) (alt. common anchor, admiralty plan)
- Admiralty pattern (modified – single-fluke mooring anchor)

- Baldt anchor
- Bow anchor
- Bruce anchor (FFTS) (FFTS Mk 4)
- Bruce cast steel anchor (Mk 2 and 3)
- Bruce Mk 3 & 4 denla anchors
- Bruce twin shank anchor
- Byers anchor
- Close stowing anchor
- Clump anchor (concrete)
- Creep anchor
- CQR (USA – plough anchor)
- Danforth anchor (close stowing stocked anchor)
- Denla (Bruce) – drag embedment anchor
- D'Hone anchor
- Eells anchor
- Flipper 'delta' anchor
- Gruson anchor
- Hall anchor
- Heavy-cast 'clump' anchor (iron)
- Herreshoff anchor (moveable stock)
- Heuss special anchor
- Improved mooring anchor
- Jury anchor (improvised emergency anchor)
- Kedge anchor (stern) (smallest anchor on board)
- Klipp anchor
- Meon Mark 3 anchor
- Mushroom anchor
- Northill anchor (seaplanes)
- Piggy anchor (alt. back anchor)
- Pool – N anchor
- Pool – TW anchor
- Sheet anchor (spare) (obsol. best anchor)
- Single-fluke anchor
- Spek anchor
- Stato anchor
- Stem anchor
- Stevin
- Stocked-close stowing anchor
- Stockless anchor (double fluked)
- Stokes anchor
- Stream anchor (alt. stern anchor)
- Tombstone anchor
- Trotman anchor
- Umsteckdraggen (four-arm/fluke anchor)
- Union anchor

Anchor Size Comparison

Figure 3.9 Bruce twin-shank anchor (18.00 kg); large, heavy-duty offshore mooring anchor. Extensively employed in the oil and gas marine production areas of the world. Comparison of size against the man standing below the anchor shackle.

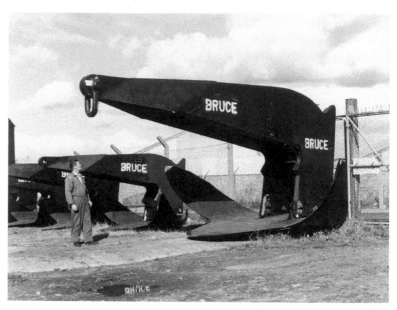

Alternative Use of Anchors

Many incidents of grounding and beaching have involved the use of anchors being deployed. Often the use of an anchor astern of a beached or aground vessel can become useful in helping to drag the ship stern first towards deeper water. Some ships carry a stern anchor for just this purpose of 'kedging' and generating an astern movement. Ships trading up to the Great Lakes of Canada were often equipped with stern anchors to prevent the stern from swinging in the wind while at anchor.

Anchors may also be an option should the vessel be caught off a 'lee shore', being used to hold the vessel off rocks, possibly following a loss of power. In such cases losing the anchor becomes possible, but the loss of an anchor compared to the loss of the ship would seem a viable alternative to most Masters.

The use of anchors is generally as a means of retaining the ship in the same position. As in running aground, the anchors are normally walked back to prevent the vessel from accidentally refloating itself on a rising tide. Using auxiliary craft to carry out a ship's anchors is not unknown. Such action would, however, require a substantial craft to carry a 15+ tonne anchor. In the days of sail, the ship's lifeboats had the capability to carry out a light anchor. However, in this day and age, with commercial ships being equipped with heavier and larger anchors, deployment by lifeboats would not be a practical proposition.

Offshore drilling installations have often employed multiple anchors in a pattern to hold the rig in a defined position. Use of eight or more anchors would not be considered unusual. The main holding anchors are frequently backed up by 'piggy anchors', a second anchor being laid in tandem, for additional support against unwanted movement.

Extensive anchor patterns tend to cover a large area and engage anchor-handling vessels to carry out the task. Marker and recovery buoys are deployed with each anchor to ensure recovery once the installation moves on. General shipping would be expected to keep clear of sea areas where installations are establishing a new operation or development. The use of a ship's own anchors if inside such an area would be problematic and could lead to fouled anchors and/or cables.

The Anchor Plan

Prior to use of any anchor it is anticipated that the Master would draw up an anchor plan. This should be established between the interested parties, namely, the ship's Master/Captain, the officer in charge of the anchor party and the engine room. Relevant bridge team members must also be made aware of the intention to reach the anchorage safely.

In the construction of any anchor plan, the following items must be worthy of consideration:

1 The intended position of anchoring of the vessel.
2 The available swinging room at the intended position.
3 The depth of water at the position, at both high and low water times.
4 That the defined position is clear of through traffic.
5 That a reasonable degree of shelter is provided at the intended position.
6 The holding ground for the anchor is good and will not lend to 'dragging' anchor.
7 The position as charted is free of any underwater obstructions.
8 The greatest rate of current in the intended area of the anchorage.
9 The arrival draught of the vessel in comparison with the lowest depth to ensure adequate under-keel clearance.
10 The choice of anchor(s) to be used.
11 Whether to go to 'single anchor' or an alternative mooring.
12 The position of the anchor at point of release.
13 The amount of cable to pay out (scope based on several variables).
14 The ship's course of approach towards the anchorage position.
15 The ship's speed of approach towards the anchorage position.
16 Defined positions of stopping engines and operating astern propulsion (single anchor operation).
17 Position monitoring systems confirmed.

18 State of tide ebb/flood determined for the time of anchoring.
19 Weather forecast obtained prior to closing the anchorage.
20 Time to engage manual steering established.

When anchoring the vessel it would be usual practice to have communications by way of anchor signals prepared for day and/or night scenarios. Port and harbour authorities may also need to be kept informed if the anchorage is inside harbour limits or inside national waters.

NB. Masters or Officers in Charge should consider that taking the vessel into an anchorage must be considered a bridge team operation.

Deep Water Anchoring

Mariners tend to accept that vessels with a deep draught will on occasions be forced into an anchoring situation in deep water. Generally, Masters of such vessels do not welcome putting heavy anchors down at any time, and especially not in deep waters.

Where it is required to use anchors in depths above the norm it is very often an exceptional case like a lee shore scenario, where an anchor walked back could hold the vessel off the shoreline and prevent the loss of the ship. Strain on the cable could be exceptional, but based on the fact that loss of the anchor is better than loss of the vessel, a deployed anchor might buy additional time.

Where deep water anchoring becomes a necessity, the anchor should not be let go, but walked back all the way. Although braking systems can be notably efficient, holding the weight and cable length of a released anchor will be a monumental task. If the anchor is let go, it can be expected to gain momentum and the brake lining would probably be burnt out trying to stop the running cable.

With ever larger ships carrying anchors in excess of 20 tonnes, braking systems have been improved, but they cannot do the impossible. The acceleration gained by the anchor and the weight of cable would exceed the brake holding power of the windlass. Such action could leave the windlass damaged and out of commission at a time when the ship might need all its resources in a working capacity.

Dragging Anchors

It is essential that once the vessel is anchored an effective anchor watch is maintained. The prime function of any watch keeper is to not only keep a lookout by all available means, but also in this case ensure that the ship's position is monitored and the vessel is not dragging its anchor(s). The Watch Officer would be expected to maintain the watch from the navigation bridge and check on the ship's position by use of a minimum of three crossed visual anchor bearings. A secondary system by radar, employing the variable range marker on a fixed land object, would also be considered as a reliable indicator that the ship's position was holding steady. Indication of dragging ship's anchors could also be gained from changing coordinates on the GPS and/or feeling excessive vibration on the anchor cable. (The OOW should not leave the navigation bridge unattended should the need arise to check vibration on the anchor cable.)

Observation of deteriorating weather conditions in the region of the anchored vessel would tend to alert the Watch Officer to any ship movement and cause precautionary actions to be taken. Such options of paying out more cable or deploying a second anchor to stop the effects of dragging would be considered as possible options, open to the Master.

In the event that the vessel continues to drag her anchor(s), it would be considered prudent action to recover all anchors and cables and move to a less exposed anchor position, preferably a more sheltered anchorage with better holding ground. Mud or clay are considered the very best holding grounds for deployed anchors.

Laying at Anchor

Figure 3.10 Vessels lay at anchor off Gibraltar in the Mediterranean. Sufficient distances between vessels must provide adequate swinging room when vessels swing due to changing weather and sea conditions. The more exposed the anchorage the higher the risk of dragging the anchors.

Another Vessel Dragging Towards Your Own Ship

The majority of commercial ports have designated anchorage sites which tend to attract increased traffic. The close proximity of several vessels in an anchorage is one of the drawbacks associated with the need for ships to go to anchor. In the event that another vessel starts to drag its anchor and is seen moving towards your position, it must be considered as a potentially hazardous situation.

In such an event, the Officer of the Watch should immediately inform the Master of the vessel and sound five or more short and rapid blasts on the ship's whistle. VHF contact with station identification would also be considered appropriate to clarify future actions to avoid the two vessels closing.

The Master has limited options to protect his own vessel from an oncoming ship, because any movement of his own vessel is hampered by his own anchors and cables being deployed. It should be realised that when the vessel actually anchors, the ship's engines should have been left on 'stop' (not finished with engines). As such, the vessel could be taken out of harm's way by steaming over the ship's own anchor cable.

An alternative would possibly be to give your own vessel a sheer away from the line of approach. This can be achieved by going hard over on the wheel so that the vessel will sheer away from a pivot point in the bows. This sheering action is possible because of the stream of water which bypasses the rudder in the aft position. If a stream of water is passing the rudder, the rudder remains effective, allowing the ship to sheer away out of the line of approach and away from immediate collision danger. The ability for any ship to sheer away from a danger is limited at best and would take time to achieve without any guarantee of success. Specific weather conditions could be expected to influence a positive outcome in any event.

Many ships are now fitted with bow and stern thruster units. Where these are available in a situation where another vessel is dragging its anchor towards one's own vessel, thruster(s) could become a useful tool to evade contact. Prudent use of thruster units could be used to physically move the vessel sideways or angle the hull away from the oncoming danger, particularly a stern thruster. (Bearing in mind the pivot point for a vessel at anchor falls right foreword, in the bow.) The use of a stern thruster could cause an influential angle to align the hull away from an oncoming danger, as with a ship dragging down.

The successful use of thrusters is limited, especially if they are tunnel thrusters, which only provide opposing directional thrust. If they are of the 360° azipod type, these can be better engaged to give a variable directional movement to the hull.

Use of Stern Anchors

Many vessels are equipped with stern anchors, but current practice would seem to reflect that they are rarely, if ever, employed during the ship's lifetime. The trade and experience of the Master are both factors that largely influence the use of a stern anchor, if carried. This situation has existed for some time and has resulted in fewer mariners having any experience at all with stern anchor operations. Nor do the Classification Societies specify any guidelines for the design or operation of stern anchors.

Vessels that have a stern anchor fitted do not normally carry a large quantity of chain cable with it, usually up to about three shackles of chain, together with an unspecified length of wire anchor warp. As such, it could be employed as a 'kedge' anchor in the event of the ship grounding. The anchor itself would be carried out and placed by a tug or other similar craft, bearing in mind that stern anchors are generally much smaller and lighter than the working bower anchors.

They are virtually always fitted on the fore and aft line, with a central single 'hawse pipe' and operated by a single cable lifter/cable holder or small single windlass. Other operating systems may include a combined capstan incorporating a cable lifter with a vertical mooring drum.

Where salvage operations are concerned, the task of laying out anchors astern is a more regular activity, especially where a vessel has run aground and an attempt to refloat is expected to take place. Anchors deliberately laid in this manner are often referred to as 'ground tackle'. They are laid with the object of keeping the vessel stationary or providing the means for the vessel to heave itself off a shoal, assuming that the watertight integrity of the hull is not infringed.

It would be considered impractical for the larger anchors of modern vessels to be carried out by ship's boats and in virtually every case of laying out anchors, or placing ground tackle, a tug or other similar craft would need to be employed.

Use of Reaction Anchors

Reaction anchors are extensively employed in salvage operations and their prime function is to reduce adverse movement of the vessel being salved. In the case of the example shown below, they are used to prevent the recovered vessel moving towards the dock when righting is effected. The act of righting the vessel will in itself take mammoth forces and the momentum generated could over-carry the vessel towards a further danger or even roll the vessel through an arc of 360° and render the ship back to her initial position.

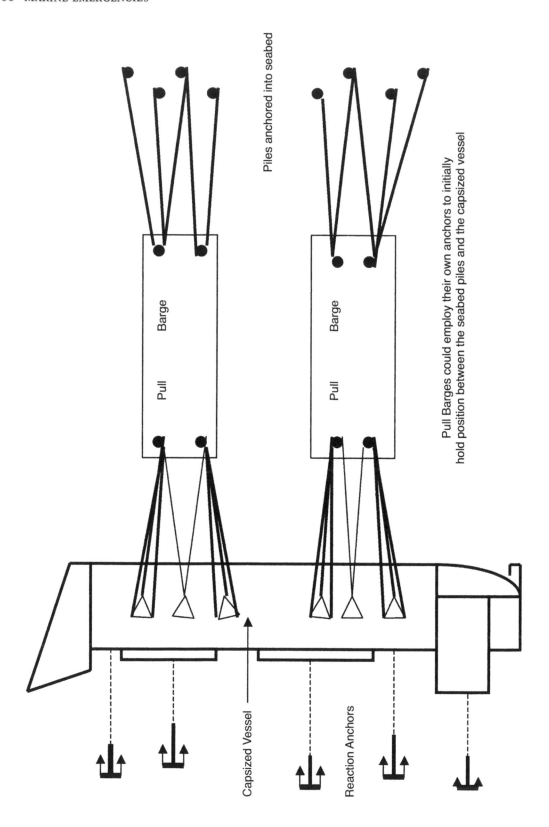

Figure 3.11 Plan view of righting operation for capsized vessel in open waters.

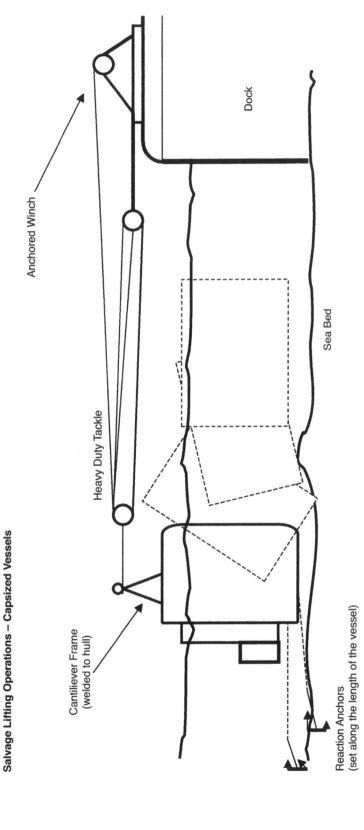

Salvage Lifting Operations – Capsized Vessels

Anchored Winch

Dock

Heavy Duty Tackle

Sea Bed

Cantiliever Frame
(welded to hull)

Reaction Anchors
(set along the length of the vessel)

Figure 3.12 Winches are anchored into the dockside and cantilever frames are welded onto the side of the capsized vessels hull. This pulling power is established the whole of the ship's length. Where a dock side wall is not part of the geography then barge platforms can be anchored in position to provide a platform for leverage in open waters. Reaction anchors are laid prior to the commencement of the operation, to prevent the vessel from rolling through the vertical and capsizing on the opposite side.

Reaction anchors are employed to reduce over-carry and momentum and are used extensively down the ship's length, with the view to reducing movement of the salved vessel. The type of anchors used as reaction anchors vary, but the Stevin 9–10 tonne anchors are favoured as first anchors and these may be supported by back-up, 'piggy' anchors. (Stevin anchors are high holding power and are constructed with a wide-set fluke arm, providing ideal qualities for such an operation.)

As with many sub-surface operations, divers are often employed with salvage operations and it would be usual for anchors to be inspected once bedded in, prior to commencing any righting activity.

A similar principle is employed with 'drag chains' when launching a new vessel in order to curtail the ship's momentum when clearing the slip. Although anchors in this case are not employed because minimum ship's movement is desirable and anchors could possibly restrict total movement.

If reaction anchors are not used, some other alternative, such as 'piles' may be placed to provide an offshore anchorage position. The nature of the holding ground would greatly influence the type of operation that would eventually dictate the method of incurring the least damage at the least cost.

Winches are anchored into the dockside and cantilever frames are welded onto the side of the capsized vessel's hull. This pulling power is established down the whole of the ship's length. Where a dock side wall is not part of the geography, barge platforms can be anchored into position to provide a platform for leverage in open waters. Reaction anchors are laid prior to the commencement of the operation to prevent the vessel rolling through the vertical and capsizing on the opposite side.

Example Stern/Kedge Anchor

Figure 3.13 An admiralty pattern anchor clamped in the vertical against an aft position bulkhead, for use as a kedge anchor.

Kedge Anchor

The term 'kedge anchor' is an expression which is given to any anchor which is engaged, or could be engaged, in 'kedging', namely the dragging of the ship, usually astern, off a shoal. It is now considered an uncommon practice, except possibly in salvage work, or when the need is extreme. It was usually achieved by carrying out and laying an anchor by use of the ship's lifeboats, but the increased weights in anchors for today's larger ships would most certainly make the practice unsafe and impractical in this modern age.

NB. Historically, in the days of sailing ships, the vessel's boats could carry out the ship's wooden anchor, say after the vessel had been blown down onto a shoal, but with the advances in industry and more specifically steel manufacture, the heavy iron/forged cast anchors make the practice obsolete.

With a salvage operation, the more popular alternative to 'kedging' is generally provided by the use of a tug (or tugs) when available. It would be a practical option to employ a tug to carry out an anchor, but if you have the services of the tug to pull the vessel clear of an obstruction, would there still be a need to carry out a kedge anchor?

A kedge anchor would normally be engaged with an anchor warp or a combination of chain and warp, being led to an onboard winching arrangement once an anchor is carried out by means of a tug or motorised barge of suitable size to accommodate the anchor's weight. Once laid it is often referred to as 'ground tackle', the purpose of which can be twofold: (1) to hold a vessel in position against accidental movement, or (2) to heave the vessel astern when attempting to refloat. It is worth noting that ground tackle alone is not normally totally effective for holding the larger tankers. It is generally accompanied with prudent ballasting which is considered a more effective holding arrangement with such large vessels.

Kedge anchors, if being deployed, are definitively placed and often buoyed to identify their position. They can be fitted with tripping/recovery lines depending on anchor type, weight and the capabilities of the recovery vehicle. Consideration of the nature of the holding ground prior to deployment is essential to ensure that the anchor is not 'drawn home' through the ground once the weight is applied to the anchor warp. Kedging has become dated and tugs would generally be considered the more acceptable option in this day and age.

Chain Cable/Stud Link: General Information

Volume of anchor chain – Anchor cable when stowed in a chain locker can be estimated at approximately 0.5 cubic metres per metric ton of chain.

Anchor chain renewal – Lloyd's require any length of anchor cable to be renewed if the mean diameter at its most worn part is reduced by 11 per cent below its original diameter.

Size of cable measurement – Cable is measured by use of external callipers. The size is found by measuring the diameter of the bar from which the link is made.

Chain grades – A method of indicating the quality of steel from which the cable is manufactured. The grades have been internationally accepted and are recognised by the Classification Societies, and are listed under their regulations.

U1 Mild steel chain
U2 Special-quality steel chain
U3 Extra special quality chain (Ref. Table 3A, Lloyd's Rules)

Grade three cable – The lightest of the three grades of cable; ships so equipped would expect to use an increased scope when anchoring.

Wrought iron cable – The most expensive to produce and weaker than the other three qualities of forged mild steel, cast steel or special-quality forged steel. Subsequently it is rarely seen on present-day merchant vessels. Its replacement is generally by a non-ferrous cable manufactured from an aluminium–bronze material.

Baldt or dilok cable – High-strength cable which is not easily comparable with British chain, size for size. Widely employed in US warships. The nearest equivalent is probably cast steel chain.

Shackle length – Anchor cable is universally manufactured and used in shackle lengths of 15 fathoms (90 feet) or 27.5 metres.

Strength – Studlink chain is 1.6–1.8 times stronger than the iron from which it is made. It is also 50 per cent greater than 'open link' cable.

Windlass capability – The cable lifter, when new, must be able to lift 3.5 shackles of cable in the vertical, with the anchor attached. As the equipment ages and wear and tear occurs this capability can be less than when fitted in a new condition.

Heavy Weather Encounter

Some readers may view a ship experiencing heavy weather not as an emergency. This author would disagree. Many incidents are accompanied by heavy weather conditions, like the lee shore scenario, for instance. When the sea is flat and calm and the visibility is 20 miles, things don't generally go wrong. However, it is a different story at 2 a.m. in a force 10 storm. Most vessels could expect to experience some violent movement in the ship's motions.

It is when heavy rolling or when the ship is pounding that movements and vibrations cause systems to fail. Gyroscopes topple, generators come off the board and screw race causes the governor to cut in. Even the best of sailors will feel uncomfortable. Heavy weather, if it persists, is tiring on watch keepers, and tired men make mistakes.

Heavy weather, with or without violent motions on the vessel, is reason enough for the Officer of the Watch to call the ship's Master to the bridge. Where heavy pitching is experienced, a reduction in engine speed will usually ease motions on the vessel and eliminate any pounding that the vessel is subject to. Where excessive rolling is a cause for concern, the ship's course should be altered to bring the wind off the bow if appropriate. The danger associated with synchronised rolling or pitching is always a real threat and swift action should be taken by navigators to eliminate such motions.

The concerns are that the period of encounter of a wave must be changed as soon as synchronisation is noted. To this end beam seas should be a major concern to all Watch Officers and avoided whenever possible. Masters should ensure that their junior officers are fully aware of the dangers of synchronisation, where the period of roll matches the period of encounter.

Synchronised pitching is rare but if encountered can be destroyed by a change of speed. It is especially dangerous if allowed to continue in an area of abnormal waves, such as a position southeast of the coastline of South Africa. Recent research has revealed that 'freak waves' are not as abnormal as first thought and satellite imagery has revealed freak waves in waters around New Zealand and in the North Sea.

The objective of ship handling in heavy weather is to try to reduce incurred ship stress and eliminate the possibility of shipboard damage. This is not an easy task against a million tonnes of water in a single wave. Riding a ship in heavy seas is exhilarating, but dangerous and requires expert control of course and speed. Experience will be gained on every occasion, but there is always more to learn.

Figure 3.14 The *British Liberty*, a BP tanker, seen in heavy seas abeam of Barrett's Reef, in the main channel on approach to Wellington, New Zealand. My thanks to Peter M. Stacey (marine pilot) for his kind permission to use this image.

Heavy Weather Preparations

Where any vessel, deep sea or coastal, is expecting to encounter heavy weather, the ship's Chief Officer would be expected to make any and all preparations to secure the ship's decks to ensure a continued safe passage. Such actions could expect to include any or all of the following.

Stability

- Improve the ship's stability (GM) by adding ballast.
- Remove any free surface moments in tanks where possible.
- Close all watertight doors.
- Pump out any swimming pools.

Cargo

- Tighten up lashings on all cargo parcels, especially deck cargoes.
- Add additional lashings and securings to deck cargoes.

Navigation

- Secure the ship's navigation bridge against adverse rolling of the vessel.
- Obtain updated weather forecast.
- Plot storm position continually.
- Update ship's position and inform shore-side authority.
- Revise ETA.
- Consider re-routing to avoid bad weather areas.
- Engage manual steering.

Deck Security

- Check all scuppers are clear and freeing ports are free to shed slack water.
- Close off deck ventilation apertures.
- Rig lifelines fore and aft to ensure safe passage through the vessel.
- Tighten up securings on survival craft and life saving appliances.
- Remove manpower on deck and operate heavy weather routine.
- Remove any awnings.
- Clear decks of any loose or surplus equipment.
- Check anchor securings are tight and spurling pipes covered.
- Close weather deck doors.
- Warn heads of departments of impending heavy weather.
- Secure all derricks/cranes against unwanted movement.
- Ensure lashings on the accommodation ladders are secure.
- Organise meal reliefs if required.
- Operate all crew members on heavy weather work routine.
- Note all preparations in log book.

Most companies would operate a checklist to use for ensuring all precautions for heavy weather are followed.

Beaufort Wind Scale and Sea Description

Table 3.1

Beaufort Scale Number	Wind Description	Wave Description	Height of Sea (Metres)	Speed (Knots)
0	Calm	Flat, calm, mirror smooth	–	0–1
1	Light air	Small wavelets, without crests	0. 08	1–3
2	Light breeze	Small wavelets, crests glassy but not breaking	0.15	4–6
3	Light breeze	Large wavelets, crests beginning to break	0.61	7–10
4	Moderate breeze	Moderate waves, becoming longer crests, breaking frequently (white horses)	1.04	11–16
5	Fresh breeze	Moderate waves, longer with crest breaking (many white horses)	1.83	17–21
6	Strong breeze	Large waves forming, crests breaking more frequently	2.87	22–27
7	Strong wind	Large waves, streaky foam	4.01	28–33
8	Gale	High waves, increasing in length, continuous streaking of crests	5.49	34–40
9	Strong gale	High waves, crests rolling over, dense streaking	7.0	41–47
10	Storm	Very high waves, overhanging crests, surface white with foam	8.84	48–55
11	Violent storm	Exceptionally high waves, surface completely covered with foam	11.27	56–65
12	Hurricane	Air filled with spray, visibility impaired	–	Over 65

NB. Safety message for tropical storm or winds of force 10 or above, for which no storm warning has been received. The Master of a ship which encounters such conditions is bound to communicate relevant information by all means available at his disposal to other ships in the vicinity and also to the competent authorities. The form in which the information is transmitted is not obligatory. It may be in plain language (preferably English) or by means of the international code. (Reference SOLAS Chapter V – Safety of Navigation, Regulation 31.)

Navigation in Heavy Weather

Planning for any sea passage is carried out well in advance of executing the voyage. Even with long-range weather forecasts the state and quality of weather conditions cannot be guaranteed in advance. However, it can be anticipated at the planning stage and adequate margins of safety can be incorporated at focal points, known landfalls and associated hazardous areas.

Waypoints off landmarks should take account of the season and the prevailing weather at that time in the geographic regions. Summer periods, usually associated with better weather than winter conditions, should influence recommended passing distances off landmarks. Eight miles in summer may be acceptable, but 12 miles in winter may be a safer alternative.

Every passage plan must have built-in contingences, and where heavy weather is predicted and expected to be encountered, the possibility of re-routing must be a consideration for the ship's safety. Just as altering course away from known ice positions would be expected, changing course to avoid heavy weather conditions is frequently a viable option.

Where storm positions are known to exist, the Masters of ships en route would be expected to plot any movement of the storm. Course alterations could ensure that the ship passes clear of the immediate threat of bad weather. Gale and storm conditions are always a potential threat to the wellbeing of cargo and crew alike. Certain cargoes are liable to shift if the vessel experiences heavy rolling conditions. Such a movement in cargo parcels could clearly have a detrimental affect on the ship's stability, so escalating the safety problems of the vessel.

An alternative route, which may not always be possible, must be seen as prudent action to ensure the continued safety of the voyage. Ship damage or cargo damage will without doubt incur insurance claims and any action to avoid bad weather is then seen as cost-effective. Severe weather conditions are an obvious threat to life at sea and in compliance with the first principle of Safety of

Life at Sea, being paramount, any action away from a danger is to be applauded.

Ship Handling in Heavy Weather

Violent movements by a ship are dangerous and uncomfortable. Any vessel which encounters the heavy weather conditions associated with gales or storms must handle the vessel to ensure the safety of all on board. Avoidance of bad weather in the first place is by far the better option, but where this is not possible and the vessel finds herself faced with such adverse conditions, changing the ship's speed and course are ways to ease the vessel's motions.

Once encountered, the vessel might be inclined to either roll more or pitch more, depending on the wind/sea direction relative to the ship's fore and aft line. Pitching is generally preferred to rolling and altering the ship's course to position the wind one or two points off the bow should give rise to a pitching motion rather than a rolling action.

In very heavy seas pitching could be expected to be violent, causing the vessel to slam into the surface, known as 'pounding'. This cannot be allowed to continue for any length of time as the vessel would absorb structural damage. To reduce and hopefully eliminate pounding, the speed should be eased.

Reducing the ship's speed when pitching has many advantages for the ship. The period of encounter with the waves is changed and the ship is not pushed into a slamming action with the surface water. The other benefits from this action of reducing speed will make structural damage less likely, crew will be more comfortable and there would be less chance of 'screw race'.

Where a vessel is pitched downwards in the bow region, the aft end with the propellers is frequently thrown upwards with opposing motion, allowing propellers to break the surface. Such action would allow screw race and cause the governor on the engines to cut in, so possibly stopping the engines. The action of the governor is to limit engine revolutions, overheating and machinery damage being inflicted. Overtaxing the vessel's machinery is never a good idea as this can only lead to an eventual loss of power and leave the ship at the mercy of the weather.

There may be an occasion when a vessel may find herself in a position of being affected by a combination of bad weather and strong adverse currents. In the event of such combined forces being brought together, the possibility of also being landlocked could eliminate any re-routing option. Typical geographic examples are the Mozambique Channel, the Taiwanese Straits and the English Channel. The prevailing conditions would probably leave the ship's Master with the single option of heaving to.

In such a situation the ship's Master should not try to make headway, but should move the ship's head to meet any wind direction change, while at the same time keep enough revolutions on the engines to hold position. The bad weather will continue to move from the current area and once conditions improve the ship could then attempt to get back on track and make headway.

Case Study: The Loss of the *M.V. Braer* (89,730 dwt)

In January 1993 the oil tanker *M.V. Braer* was en route from Bergen in Norway, towards Quebec, Canada, carrying 85,000 tonnes of Norwegian 'Gullfaks' crude oil. On 3 January, while off the Shetland Islands, in an estimated position about ten miles south of Sumburgh Head, the vessel became damaged in storm force winds, caused by pipeline sections breaking free on the tanker's upper deck.

This damaged area caused a loss of the ship's watertight integrity, which allowed sea water ingress into the engine room, causing the vessel's engines to stop and the control of the ship to be lost. Attempts to restart the engines were made in vain, probably because of fuel contamination.

The vessel, although without power, was not in immediate danger, but was drifting towards the land under the influence of south-westerly winds of force 10–11. Coastguard rescue helicopters were alerted at Sumburgh and RAF Lossiemouth and it was decided to remove 14 non-essential crew members of a 34-man crew using helicopters.

At 0850 hours on 5 January the Greek Captain was persuaded to abandon ship. At the time, the vessel was experiencing strong, local north west currents, which caused her to drift, moving to miss Horse Island but towards Quendale Bay.

The anchor-handling vessel *Star Sirius* was called in to try to establish a towline, but in the extreme weather prevailing, contact was impossible. The vessel eventually grounded at Garths Ness and the remaining personnel were recovered by helicopter.

Oil Spill

From the time of grounding, the *Braer* started to spill her cargo, losing it all, some 85,000 tonnes, to the sea. Considerable damage to birdlife and mammals occurred, together with fish stocks and shellfish sources. Farmed salmon fisheries in the area were tainted by the spillage, with stocks having to be culled. Compensation was later paid to fish farmers.

Long-term environmental damage to the area was anticipated, but this was not the case, for two reasons. First, the type of oil being carried from the Norwegian Gulfaks oil field was a lighter grade of oil and more biodegradable than other North Sea oils. Second, the extreme bad weather which had hampered rescue and preventative measures being taken dispersed the oil more quickly by wave action and evaporation.

Lessons Learned

At the time of the *Braer* incident the oil spill was the eleventh largest that the world had experienced. Although the effects were not as bad as anticipated, the incident did provide a warning of the potential danger in the future.

From the shipping side, bearing in mind the weather and sea conditions were extreme, the rigging of lifelines to a ship's decks is essential to provide access in any condition to all positions aboard the vessel. This would include a position well forward, where the anchor-handling equipment is established. Men might be required to deploy anchors, but if they cannot reach the windlass controls the possibility of emergency anchor use is lost.

Also, the passage planning of a voyage is meant to be flexible in its execution. Extreme bad weather, if known to exist ahead of a ship's route, should influence course tracks with a view to avoiding poor conditions. The position of Bergen in relation to the Shetland Islands is such that a few degrees of alteration could have taken the ship north about the islands rather than south of the islands, where it was influenced by the south-westerly winds.

Of course, with hindsight decisions come easier to historians. A more real test exists on the navigation bridge when formulating all aspects of a passage plan. Ships Masters do not have an all-seeing 'crystal ball' and must make their decisions on experience and the prevailing conditions at the time.

Shortly after the loss of the *Braer*, oil tankers over 20,000 dwt had to be fitted with emergency towing arrangements under the IMO resolution MSC.35 (63).

Today, all ships over 500 grt and passenger ships must be fitted with emergency towing arrangements fore and aft, pre-rigged and capable of easy deployment.

NB. The *Braer* incident accelerated the establishment of this emergency towing standard from the Marine Safety Committee of the IMO.

4

Fire on Board

Introduction

Fire on board a vessel at sea is a most frightening situation. Mariners, the world over, continually train to combat fire. They anticipate that sooner or later fire will strike, usually when it is least expected. Fires on shore can be tackled by the local brigade, but a fire at sea, in the middle of an ocean, means the crew are very much on their own.

The 1974 SOLAS with the 1978 protocol and the amendments of 1988/1992 provide the fire regulations and requirements for the shipping tonnage of today. Design features in ship construction have been incorporated to reduce risk of fire occurring and at the same time contain any potential fire outbreak with the onboard firefighting systems. Crews conduct regular training drills and are generally well prepared to tackle any fire aboard the vessel. Unfortunately, fire is an intangible and it doesn't follow the regulations as one would expect, especially when coupled with the element of human error.

Different types of fire require differing treatments; for example, water as a medium is good for carbon fires, but dangerous for oil or electrical fires. Shore-side statutory training courses go some way to develop seafarers in the methods and approaches to firefighting various fire types, but we still see mistakes being made.

The additional burden with fires at sea are, of course, related to the fact that the vessel does not provide a stable platform. Excessive use of water can generate free surface moments and cause the ship to become unstable, as happened to the *Empress of Canada* in Liverpool, England. In January 1953 the ship was destroyed by fire in Gladstone Dock; the vessel subsequently capsized from the weight of water employed from firemen's hoses when tackling the fire. It took until March 1954 to complete the salvage of the vessel.

This example might be considered as a fire in port, where the weather was not seen as a critical factor. A ship at sea with fire on board can expect to be more exposed to the adverse elements of inclement weather. Firefighters can expect to find great difficulty on wet, slippery, angled decks due to the rolling and oscillations affecting the stability of the hull.

The nature of shipping is involved with a variety of cargoes from bulk crude oil to containers carrying everything from the kitchen sink onwards. Hazardous goods are listed under the International Maritime Dangerous Goods code (IMDG), used throughout the marine industry. If such items are encompassed within a fire at sea, toxics and explosions may occur, in addition to burns and smoke inhalation. Our seafarers must be prepared for the worst of all outcomes.

The Outbreak of Fire on Board the Ship

Although there are several categories of fire with various location possibilities aboard the vessel, there are basically only two fire scenarios: (1) fire at sea; and (2) fire in port.

In every case, without exception, whoever discovers a fire outbreak should raise the fire alarm. The action of raising the fire alarm would be expected to put the vessel on an alert status, allowing fire parties to be mustered and firefighting to get under way.

On the premise that all large fires start from small fires, training should reflect that whenever possible the fire should be tackled 'immediately' with whatever is readily available in an attempt to curtail spread. Such action should not be foolhardy by any individual, but should be positive action after the alarm has been sounded.

Effective training and on board drills should generate rapid response times to the sound of the initial alarm, namely continuous ringing of the ship's fire bell(s). Well practised crews would be expected to move towards designated fire stations and make ready respective equipment as they move towards a fire scene.

Fire at Sea

Once the alarm has been activated the ship's Master would be expected to move to the navigation bridge and take over the 'conn' of the vessel. It must be anticipated that he would order the immediate position of the ship to be placed on the chart. Simultaneous actions would also include: closing all watertight/fire doors, obtaining a weather forecast, changing from auto steering to manual steering once a helmsman reports to the bridge, displaying NUC signals and advising the nearest coast radio station.

Preparation of an URGENCY message or a MAYDAY communication would be expected. The context of this would be influenced directly by the firefighting actions of the crew. Where a fire escalates out of control and cannot be held in check or extinguished easily, the Master could expect to order survival craft to be turned out and made ready for immediate launch.

Where a passenger vessel is involved, a prudent action might be considered to load passengers into lifeboats but hold off the launch until the vessel becomes no longer tenable. Early movement of passengers could be seen as precautionary in order to avoid panic.

Personnel aboard passenger vessels are trained in 'crowd control' but in reality one cannot expect the circumstances to remain calm when tensions are running high.

Fire in Port

Fire aboard a ship in port is just as serious as a fire at sea because it has its own inherent dangers associated with the close proximity of port facilities. In every case of fire the fire alarm must be activated to alert all personnel to the closeness of the immediate danger. Personnel on board could be expected to include shore personnel such as stevedores (longshoremen), ships' agents and visitors to the vessel, customs officers, security personnel, auditors and cargo surveyors.

The ship's own gangway security system would be expected to monitor all visitors to the ship who are on board at any one time. Similarly, they would expect to be aware of any of the ship's crew who are ashore at that moment in time. With variable numbers of personnel on board the vessel, the Chief Officer would be expected to act positively in the event of an outbreak of fire:

- *Immediate call to the local fire brigade* (probably via the ship's VHF to the Port Control).
- *Send all unnecessary personnel ashore* (to reduce the immediate threat of loss of life aboard the vessel).
- *Place a Chief Officer's messenger at the head of the gangway.* He would be expected to move to take charge of the fire parties and attack the fire scene immediately.

Clearly, the location of the fire will influence future and subsequent actions. The presence of the Chief Officer's messenger at the head of the gangway is to meet the shore-side Fire Brigade Officer when he arrives at the ship.

> **NB.** The controlling authority for the incident remains the ship's Master or his designated agent. The responsibility is not handed over to the local Fire Brigade.

Clearly it is in the interests of all to work together to extinguish the fire, but the overall responsibility for the fire on board the ship remains in the hands of the Master.

It should be realised from the outset that many developed countries around the world have excellent emergency fire defence services. However, many under-developed countries have little in the way of experience and resources to tackle a major fire aboard ship. This is especially so in smaller ports of South and Central America and around the Asian coastlines.

With this in mind ships' crews may find themselves reliant on their own firefighting facilities aboard their own ships and not having any real support from shore-side fire brigade systems.

Fire Support Units

Figure 4.1 The effects of fire at sea can be extremely destructive; ports and offshore operations have established support units to be actively engaged in firefighting operations from a secondary platform. The *Pacific Retriever*, an anchor-handling vessel, is seen displaying her firefighting capabilities.

Fire Parties

Provided crew have been drilled and trained correctly in firefighting practice, it must be assumed that personnel will be influential in a particular duty. The fire parties will be actively engaged in the following duties:

- *Ventilation party*:
  - This party could expect to be engaged in closing off the oxygen to any spaces where fire is discovered. Funnel fiddley above engine room spaces generally have closure covers and the funnel itself will usually have closure flap arrangements to cut off normal air supplies.
  - Cargo holds and cowl ventilators can be removed and plugged while mushroom ventilators are frequently covered to stop air flow into cargo spaces.
  - Air pipes to lower compartments are sealed by plugs and canvas covers to generate a sealed atmosphere (so removing the oxygen content of the fire triangle).
- *First aid party*: usually on stand-by with stretcher and medical equipment to deal with any casualties. Prepared to treat for burns and smoke inhalation.

- *Damage control party*: actively engaged in boundary cooling on as many of the six sides of the fire as is possible. Boundary cooling can expect to be continuous even after the fire is thought to be extinguished.
- *Fire party*: with breathing apparatus at the fire face with an active firefighting medium. Hose branch lines with water or foam compound, depending on the nature and category of fire.
- *Bridge team personnel*: Master on the conn, helmsman, lookouts, Officer of the Watch.
- *Communications*:
  - Efficient communications should ensure that all fire and damage control parties work together to close the incident down. The ship's Chief Officer could expect to be the lynch pin with the majority of fire incidents on board.
  - A three-way communication link should be conducted between the navigation bridge, the Chief Officer himself and the fire face team.

Chief Officers, through the communication link, could expect to be in charge of logistics, such as: the supply of foam compound, establishing water pressure on fire mains, re-supply of manpower, maintaining air supplies to B/A bottles, keeping the Master advised and monitoring of progress of firefighters.

It would clearly be the role of the Chief Engineer to recommend to the Master as to the use of CO_2, especially in the case of machinery space fires, bearing in mind that the use of total flooding with the inert gas would lead to a dead ship, without engines.

Firefighting Teams

Fixed firefighting installations are a statutory requirement for most ships' engine rooms.

Steam smothering systems can still be found in some smaller vessels, but in the main most vessels carry either a bulk or bottle bank CO_2 system, of which the bottle bank contained in a designated CO_2 room is by far the more popular. More recent construction has moved into a compartmentalised operation, where a calculated number of bottles are earmarked for a designated protected space, with other bottles allocated to protect other spaces.

In every case the number of bottles to be deployed in the event of a fire, in a protected compartment, will be defined by the ship's CO_2 plan. These plans are displayed in the CO_2 room itself, emergency control rooms and/or the Chief Mate's office. An inset miniature diagram is usually included on the vessel's general firefighting arrangement plan.

Example CO$_2$ Bottle Bank System

Figure 4.2 System operation activated by remote cabinet. Alternative operation can be made from the CO$_2$ room itself. CO$_2$ rooms are treated as an enclosed space.

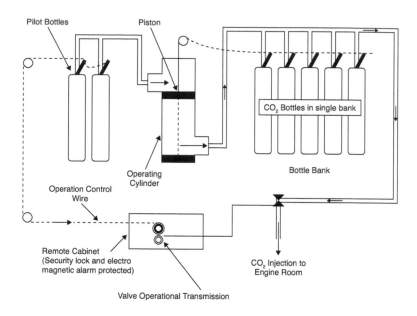

CO$_2$ Maintenance

The fixed firefighting system is regularly checked when in dry dock. The levels of carbon dioxide inside the bottles are tested to ensure no leakage has taken place and that the bottles are at full capacity. Checks are carried out by ultra-sonic indicators and by weight assessment. Pipework would be visually inspected and can be blown through.

Alarm circuits, both audible and visual alarms, to protected spaces are tested. Gate valves are also checked during routine docking inspections of the system.

Testing of CO$_2$ systems will generally be accompanied by associated testing of smoke detectors and/or sprinkler operations, often in the presence of a Marine Authority surveyor as part of the Safety Equipment Certificate Survey, renewed every five years with annual inspection.

NB. CO$_2$ rooms are considered as enclosed spaces. As such, personnel should enter under an enclosed space entry permit to work, in association with a risk assessment.

FIRE ON BOARD 87

Security against Accidental Release of CO_2

The total flood carbon dioxide system generally operates from either the CO_2 room itself or from a remote discharge cabinet. The remote position is usually in a location like an emergency control room or Chief Engineer's office. The cabinet door is key locked and the key is kept in a break-glass receptacle next to the cabinet door.

A second security measure is in way of a 'gate valve' which closes off the delivery line. This is kept in the closed position and would prevent the gas entering the protected space. On activation of the pilot bottles, this gate valve is opened at the same time as the pilot bottles.

Example Fires

On the basis that large fires inevitably start from smaller fires, it is considered worthwhile to look at a variety of special shipboard fire situations:

- engine room fire (high-risk area)
- cargo hold fire
- ro-ro, vehicle/deck fire
- cabin/accommodation fire
- galley fire (high-risk area)
- deck fire
- paint room fire
- fire aboard in dry dock
- passenger ship fires
- tanker vessel fires.

Engine Room Fires

The machinery space, by its very function, is a high-risk area of the vessel. Not only are highly flammable liquids associated with the space, but heat generation is an active ingredient of its working operation. Such a space brings together all three basic elements of the fire triangle: heat, fuel and oxygen.

Any fire, no matter how small, must cause the generation of the fire alarm immediately. The Master would expect to move directly to the ship's bridge and take the navigational 'conn'. Providing sea room allows, he would probably expect to turn the vessel's head, to go stern to the wind and adjust the ship's speed. This would effectively reduce the draught within the vessel and hopefully reduce oxygen content to the space affected by fire.

One must anticipate that the Master would be given a report on the size of the fire and the possibility of tackling the outbreak by conventional firefighting means.

> **NB.** Conventional means: hose branch lines, extinguishers, etc., as opposed to total flood systems for protected spaces.

Past experiences have shown that a fire of any size in this space would cause the total flood CO_2 system to be activated sooner rather than later. Once this occurs the Master would expect the engine room to be closed up and he/she would have a dead ship, without the use of engines.

Procedure to Activate the Total Flood System

1 Shut down: boilers, fan/ventilation and fuel, prior to evacuation of the engine room.
2 A head count, roll call or muster must be made to ensure the engine room space is empty of personnel prior to CO_2 injection.
3 Activate the total flood system from the CO_2 room or from the remote cabinet.

Fire parties and damage control parties would be expected to continue boundary cooling on as many of the six sides of the fire as is practical.

Use of Conventional Firefighting Equipment

Using hoses and branch lines in such a space is difficult and if the decision to engage with conventional firefighting equipment is made, then breathing apparatus for firefighters is essential.

A three-way communication link would be required between the firefighting team, the ship's Chief Officer and the navigation bridge. The role of the Chief Officer would be logistical, to supply the firefighting medium, like foam compound, and maintain fighters with an adequate air supply.

Updates to the Master could be expected to influence external communications and any decision to employ total flood would be at the discretion of the Master. Ship's Masters must be conscious that a dead ship will result from employing total flood CO_2 and leave the vessel at the mercy of current prevailing weather, tidal influence and current movements.

After Total Flood CO_2 Use

CO_2 is considered very effective as a smothering agent, but the heat from any fire will probably remain for some time. Also to be considered is the fact that CO_2, in the main, is a one-shot system and opening up the engine room too soon would allow oxygen ingress to the space and run the risk of 'flash over' reigniting the fire, with no CO_2 supply left.

Masters should not be tempted to inspect engine rooms too soon after CO_2 injection. The alternative option would be to monitor the space temperatures by the use of cargo thermometers down air pipes and sounding pipes. By graphing these temperature results, a declining graph would be positive evidence that the CO_2 and boundary cooling have been effective.

The cautionary action of waiting before carrying out any inspection cannot be emphasised enough. But how long can a Master afford to wait before ordering an inspection to take place?

Not an easy question, but the answer would greatly depend on the weather and the ship's current position being close to associated navigation dangers. As long as practical, with a minimum of 24 hours being probably a good starting point.

Entering the Space after CO_2 Flooding

At some stage an inspection will be needed. This would need the inspection team to be in protective suits and enter only with breathing apparatus. Entry should be made through the double-door airlock system and the team would not be expected to take firefighting equipment, their function being just to inspect the effects of CO_2.

> **NB.** At least one member of each inspection team should be familiar with the layout of the engine room.

A prudent Master would probably not take any action on the initial report from the inspection team. It would be expected that the Master would need to obtain a second opinion. This would warrant a secondary inspection team to enter the space.

In the event that both first and second opinions give indication that the fire is out, only then should the Master send in a fire team with spray hoses to cool off hot metals. Breathing apparatus must be in continuous use through these phases.

Once all bottom plates have been cooled down, ventilation should be allowed and fans should be activated if possible. Once the atmosphere has been tested and seen to be clear of CO_2 pockets, then and only then can the use of breathing apparatus be suspended. Emergency escape breathing devices (EEBDs) should still be worn by personnel in the immediate vicinity. It is anticipated that such personnel would be the ship's engineers looking to instigate repairs if at all possible, following the fire.

> **NB.** A second school of thought regarding the navigation of the vessel could be one where if the fire in question generates large volumes of smoke, then bringing the ship head to wind could decisively clear smoke volumes within the confines of the vessel. Differing circumstances dictate changing tactics.

Water Mist Systems (Ref. SOLAS – Regulation 10, Fire Fighting – Paragraph 4, Fixed Fire-Extinguishing Systems)

Fixed-pressure water spraying fire extinguishing systems are now required by passenger ships over 500 grt and cargo ships of 2,000 grt and above, where the machinery space is of category A, above 500 m³ in volume. Carriage of the pressure water system is in addition to the fixed fire-extinguishing system prescribed for the vessel.

> **NB.** Water pumps other than those required for the provision of water for serving the fire main and other firefighting systems must be installed outside of the space they are designed to protect. The arrangement means that a fire in the space or spaces protected will not put any such system out of action.

High-pressure water systems are now widely fitted aboard passenger vessels and ro-ro transports, coving their vulnerable high-profile risk areas like boiler fronts, incinerators, purifiers for heated fuel oil and hazardous areas of internal combustion machinery. Where periodically unmanned machinery spaces are employed, the systems must have both automatic and manual release operations.

Additionally, it should be noted that Machinery Spaces of Category 'A' in passenger vessels carrying more than 36 passengers must be provided with at least two suitable water-fog applicators.

There are several manufacturers that supply pressurised water mist, high-fog systems which have become popular in the maritime environment. The obvious advantage over other fixed systems like CO_2 is that firefighters would not expect to be overcome by carbon dioxide poisoning. The water pressure system may be uncomfortable, but the atmosphere within remains breathable.

The units are normally positioned in emergency control rooms or similar positions away from the actual protected space. The amount of water used under pressure is known to act more efficiently by generating a coolant mist effect. This mist provides greater coverage over a wide area and provides an effective knockdown capability on active fires. It tends not to generate slack water volumes that may affect the ship's positive stability

The high-pressure system should not be confused with a typical water sprinkler system, which tends to use a greater quantity of water in what is usually a comparatively small space. Sprinkler systems and water curtain operations, where employed, need an effective drainage system to clear slack water.

Cargo Hold Fires

The fire inside a cargo hold either at sea or in port will be directly influenced by the nature of the cargo. The success of extinguishing any fire in these compartments could well be influenced by good ship keeping. The preparation of the hold prior to loading the cargo could well be at the heart of a successful outcome.

The cleaning operation of cargo spaces and the removal of residuals tend to eliminate at least some risk of a fire outbreak. Enforcing a no-smoking policy inside holds should be pursued and smoking on any ship should be discouraged. Hold lighting should also be regularly inspected and maintained and protective grills should not be allowed to drop into disrepair.

However, the cleaning and checking of hold bilge suctions and scuppers must be considered an essential activity before cargo is loaded. In the event of a fire outbreak, holds can be flooded to extinguish flames. Inevitably the bilge suctions would need to be activated to clear away water volumes if deliberate flooding has taken place, once the fire is extinguished.

Clearly, a large volume of water inside a normally dry cargo hold would be expected to influence the ship's positive stability. To this end the pump suctions, if found to be defective, would leave the vessel with distinct problems. The use of water via hoses or from actively flooding the space would also be expected to cause cargo damage and the Master would probably have to declare an act of 'general average' as well as making a 'Note of Protest' following the vessel's arrival in the next port.

Coal Cargo

The nature of the cargo may allow the use of water as an extinguishing agent, but some cargoes like bulk coal or sulphur are such that the use of water would be extremely detrimental.

A fire in coal is notably a very 'hot' fire; if water was used at sea this would turn to steam and this in turn could pressurise the ship's hold, causing explosive damage.

Where steam can be vented to the atmosphere with hatch tops open, this steam pressure can be avoided, but keeping hatch tops open at sea is not to be recommended for retaining the ship's watertight integrity.

> **NB.** Coal in transport needs surface ventilation to remove methane gas and such ventilation would be expected to remove the immediate cause of fire generation.

Bulk Sulphur

A fire with a sulphur cargo should *not* be tackled by water. Sulphur itself is very corrosive, and if mixed with water it becomes excessively corrosive as a formulated acid. The way to tackle a sulphur fire is to smother the burning region with more sulphur, so removing the oxygen.

Most cargoes carry a fire risk element and as such most ships carry CO_2 protection in total flood systems covering protected spaces. Some exceptions prevail, e.g. bulk ore carriers carrying only non-flammable cargo are exempt from having a CO_2 system in cargo spaces.

Firefighting Inside Cargo Holds

Cargo spaces are large and for crew members to tackle a cargo hold fire is extremely difficult. Movement and the manipulation of heavy hoses around cargo parcels to reach a fire area is not an easy task and one that is prone to accidents. This is particularly so if the weather is inclement at the time and the ship is rolling or pitching heavily. Individual movement is further hampered by breathing apparatus inside an expected warm temperature. Communications may also be difficult under steel decks or around dense metallic cargoes like vehicles or containers.

The danger from toxics with mixed cargoes is ever-present, although the breathing apparatus will provide some defence against the release of noxious emissions within the enclosed spaces.

Case Incident

On 14 July 2012, the German-flagged container ship *MSC Flaminia* was evacuated after an explosion and fire which killed two crew members. The ship was reported to have extensive damage to three cargo/container holds, but the engine room, stern section, main superstructure and the forecastle had not been affected.

Seven ships went to the aid of the vessel and the 22 surviving crew members. Three injured crew members were taken to the *Azores* for medical treatment, while the remainder were taken by the tanker *DS Crown* to Falmouth, in the UK.

The fire was tackled by on board crew members and assisting support units. The vessel subsequently developed a 10° list and was taken in tow when the blaze was brought under control.

This was a notable incident that a European-flagged vessel did not receive permission from European countries and was so denied the call to a port of refuge.

Roll On-Roll Off Cargo Spaces

Designated vehicle spaces are usually fitted with a sprinkler system and/or water curtains, and if the spaces are enclosed will also be CO_2 protected. Motor-driven units will have the inherent risk of fuel, petrol or diesel in tanks, together with the potential fire risk lying within whatever cargo parcels they are carrying.

A specific metre lane grid for the stowage of ro-ro units is part of ferry operation systems, where the lanes are divided into bays. Sprinkler systems tend to protect bays, which may be separated by water curtains to prevent fire spreading from one bay to the next. It is therefore practical to ensure vehicles are correctly aligned within lanes and bays to ensure effective sprinkler coverage.

Each vehicle deck can expect to be equipped with fan systems to clear exhaust fumes, and an effective scupper drainage system to remove excessive water residuals from operation of the firefighting systems if called into operation.

Figure 4.3 Part-loaded vehicle deck of a ro-ro ferry. It shows a wide, open space with no division by athwartships bulkheads. Enclosed decks like these are fitted with extraction fans, sprinkler drenching systems (water mist), total flood CO_2 and conventional firefighting equipment (extinguishers every 40 metres). The effective drainage system is meant to eliminate water build-up, but where a major ingress occurs, as with the *Herald of Free Enterprise*, free surface will undoubtedly become a factor of consideration in such a wide, open area.

Dangers associated with these vessels may come from the accumulation of slack water not being cleared by the drainage system. Slack water building up could generate free surface moments and seriously affect the positive stability of the vessel.

Such an incident could be exasperated by the canvas or plastic sheeting that often accompanies ro-ro trucks. In the event of a vehicle catching fire, large portions of plastic or partially burned canvas could find their way into blocking the scuppers and drainage system, so reducing the effectiveness of water clearance.

NB. Modern ferry vessels are now frequently fitted with compartmentalised CO_2 coverage. Unlike a total flood, bottle bank or bulk CO_2 tank, such a system allows so many bottles to be designated to protect one space. So a more modern version takes away the 'only one shot' option and allows multiple spaces to be protected.

Figure 4.4 The Isle of Man/Liverpool operate the *Viking*, a passenger/vehicle ferry in and around the Irish Sea area of the UK. Modern, high-speed design fitted with davit-launched fast rescue craft and liferaft coverage. Vehicles enter the cargo deck through a stern door/ramp.

Cabin Accommodation Fire

Fire inside the accommodation block, either at sea or in port, is probably the most common of fire incidents aboard ship. Good housekeeping with effective training go a long way to preventing what are usually small fires. These tend to be influenced more by human error than probably any other type of fire.

The reasons for fire inside the accommodation are frequently one or more of the following: smoking in bed; overloading an electrical plug socket; using a smoothing iron; rubbish igniting in a waste basket; flammable material being placed around lights and lamps; and heat-generating units not being given adequate ventilation.

The action in the event of such a small fire must always be to immediately raise the fire alarm. Based on the historical fact that the majority of large fires start from small fires, an early alarm is considered essential.

Depending on the class of fire, it would be expected that the small fire would be attacked as soon as possible. Concern must be noted with electrical fires and carbon fires where live electrics are present. The isolation of all electrics should of course be completed before attempting to bring in water-based firefighting mediums.

The accommodation block is generally equipped with suitable and adequate numbers of fire extinguishers and hose reels. The use of any conventional firefighting equipment must always be with appropriate common sense and with the knowledge of which extinguisher should be employed for which class of fire.

Cabin Fire

The young Watch Officer completing his/her watch at sea would be expected to complete rounds through the accommodation block to detect anything untoward. The discovery of heat, smoke or flame would expect to result in the immediate sounding of the fire alarm. The activation of such an alarm would automatically be known on the navigation bridge through the smoke detector/alarm system. The policy of shipping companies may differ slightly, but in the main the discovery of smoke under a cabin door and the alarm being triggered could pose a dilemma for whoever had discovered the fire. Should they report to their muster station or should they stay at the scene of the fire and take action to attack the fire immediately?

In one sense, persons discovering fire on board are advised not to leave the scene of the fire unattended. On the other hand, remaining at the scene would leave the muster party shorthanded. This author maintains that whoever discovers the fire should stay on site and carry out the following actions.

Bang on the doors of connected cabins and raise the attention of persons that may or may not be inside. Bring extinguishers close to the fire scene. Run hoses from connected hydrants towards the identified cabin. Although these are positive actions of setting up firefighting equipment, *no water should be activated on or near the scene until the fuses for the electrics affecting the alleyway/cabin area have been drawn*. Such action would inevitably leave the area in darkness or subdued light, leaving firefighters reliant on emergency lighting torches and portable lamps.

Once the fire team is on scene, 'no attempt' should be made to open the door as this would only allow ingress of oxygen, causing possible flash over, accelerating the blaze. Instead, the crash panel at the base of the door should be pushed in, a hose jet should then be turned upwards through the crash panel opening and directed towards the deckhead of the cabin on fire. This action would cause the jet to be deflected and water would 'umbrella' inside the cabin, causing a cooling effect.

After this cooling has been effected, extract the hose, turn the nozzle to a spray action. Bring in a second hose on a pulsating jet and enter the space with the second hose behind the protective spray, to effectively kill the fire.

Figure 4.5 Example of a dry powder fire extinguisher in the accommodation location. Service dated and tagged for immediate use.

Galley Fire

The galley spaces of any vessel must always be considered as a high-risk area and likely to experience a fire scenario. Modern vessels are now fitted with a CO_2 injection system inside ducting for the galley space(s). Other vessels will still be equipped with extinguishers and fire blankets.

The type of stoves, oil or electric, will dictate the type of extinguishers which prevail in the area. Generally the correct type of extinguisher will be closest to hand, e.g. oil stoves = foam extinguishers; electric stoves = CO_2 or dry powder types.

A fire inside a galley would be cause to trigger the fire alarm and establish a firefighting party. Ventilation should be closed up and boundary cooling enacted on as many of the six sides of the fire as possible. Practical use of the fire blanket to douse burning fat containers could be extremely effective if employed quickly.

Current regulations require that separate firefighting systems are fitted to industrial deep fat fryers by way of extinguishing canopies. Also, thermostat overrides are a feature of modern-day galley equipment.

Burning liquids, if spilt, could be expected to increase the range of the fire and early action on containment must be seen as the most prudent action. Bearing in mind that oil on water will spread the fire, the use of water as the firefighting medium would not be considered an ideal choice. Water should never be used on burning fat containers like deep-fat fryers.

Galley spaces on passenger vessels are extensive and widely used by large numbers of personnel. Good housekeeping and calm operational practice will greatly reduce the risk of fire outbreak. Regular drills and training exercises can expect to benefit the efficiency of catering staff, who are generally those persons most likely to be first involved.

Modernisation in Firefighting Methods

Certain aspects of conventional firefighting have generated a need for more effective and less damaging anti-fire solutions. Electrical fires are generally fought by either CO_2 or non-insulating dry powder extinguishers. The CO_2 is usually more favourable because of the extensive clean-up required after the use of dry powder.

This practicality has fuelled research into the possibility of using inert gases to fight a wider spectrum of fires. Oil tankers have employed inert gas systems over many decades to provide specific inert spaces over high-risk fuel surfaces. Clearly, fire, which requires oxygen, cannot be established within an inert atmosphere.

New systems have emerged that use a clean agent as a fire suppressant. High-temperature facilities can be protected by alarm sensors which activate in the event of fire. The activation of the system directs a volume of inert gas to the fire surface, effectively extinguishing the fire without causing damage to the plant or machinery involved.

The systems work on the identification of where high-risk fire surfaces exist, allowing the volume of inert gas to overwhelm the volumetric target area. The suppressing gas is then released from a small reservoir pressure bottle direct to extinguish the fire position.

Systems using this clean agent fire suppression system have already been fitted in galleys to protect cooking ranges and accommodations for electrical appliance boards and computer installations.

Paint Locker Fires

Paint rooms generally contain many flammable substances, not least paint itself, in addition to turpentine, paint thinners, cleaning oils and similar accelerants. The majority of paint lockers tend to have steel doorways and are usually protected by a sprinkler water system.

Figure 4.6 Galley fire protection system CO_2 for ducting and canopy hood. Also, an inert gas system direct to high-risk deep fryer and hot plate surfaces aboard the ro-pax vessel *Clipper Point*.

The main causes of an outbreak of fire could be linked to an electrical fault in the region of the deckhead or bulkhead light. Alternatively, the cause is likely to be from spontaneous combustion where cotton waste or oil rags have been discarded and become wet.

A paint room fire must be seen as an extremely hazardous environment due to toxics given off by burning paint. Exploding paint containers and fumes from the same make such fires difficult for firefighters to approach. Breathing apparatus should be worn by in-close firefighters and boundary cooling should be considered as standard practice in tackling such an outbreak.

Early activation of the fire alarm and the sprinkler system will go a long way to containing this type of fire sooner, rather than later. If the fire is allowed to gain a hold it must be realised that it would become a very 'hot' fire and difficult to extinguish.

Automatic Sprinkler System

Many vessels are now fitted with water sprinkler systems as a fire protection/firefighting necessity, especially on passenger vessels, ro-ro ships and ferries. Systems tend to vary with manufacturers, but the basic operation elements tend to function along similar lines.

Sprinkler System Checks

Sprinkler systems are tested under the safety equipment surveys and must also be reset in the event of operation. In order to carry out the test, each section valve should be partially operated to

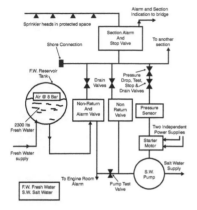

Figure 4.7 Example of automatic water sprinkler system. Many vessels are now fitted with water sprinkler systems as a fire protection/fire fighting necessity, especially on passenger vessels, Ro-Ro ships and ferries. Systems tend to vary with manufacturers but the basic operation elements tend to function along similar lines. Example water sprinkler system suitable for Vechicle Decks, of Ro-Ro vessels, Passenger cabin arrangements of specialist rooms e.g. paint rooms.

ensure that the bridge alarm/communication is operative. The test valve should then be opened, which will allow water into the pressure alarm switch. Both bridge and engine room alarms should activate. As water flows to each section, the engine room alarm should sound.

In order to test the salt water (SW) pump, close the drain valve. A build-up of pressure from closing the drain will be experienced, indicating the pump is functional. After testing is complete the system must be re-pressurised with fresh water and air, making sure that the pressures are the same on either side of the valves.

After Activation

Switch off the section valve affected. Replace the sprinkler head(s) bulbs which have activated. Re-pressurise the system with fresh water and air, ensuring the pressures either side of the valves are the same.

Maintenance

Sprinkler systems tend to be inspected and tested at dry dock periods. This would generally include a piping and valve inspection throughout, as well as a test on the system. The system is monitored under planned maintenance and would be liable to a detailed inspection for the safety equipment survey of the vessel's life saving appliances.

Fire Aboard in Dry Dock

The dry dock scenario is probably one of the highest fire risk arenas that a vessel can find herself in. Ships undergo refits and repairs during routine and emergency dry docking periods. It is during these periods that the vessel is exposed to extensive hot work in the form of steel cutting, welding and burning.

The element missing in the event of fire in a dry dock is of course the lack of water, which is always available when the ship is at sea. In dry dock it is exactly that – a *dry* dock, free of water around the hull. This is not to say that the fire water-main is not pressurised. The ship's Chief Officer and the dry dock management would both want the fire water-main pressurised as soon after docking as possible.

The probability of a fire outbreak while in the dock exists, and just like a fire in port it must be anticipated that the local Fire Brigade would be called in to deal with any outbreak. The dry dock itself may have its own firefighting units. However, this does not relieve the ship on docking of the requirement to have basic utilities in place, like telephone landline communications in order to bring in local emergency services. Some might say that this is unnecessary in the days of the mobile phone. This author would disagree. Mobile phones have the added risk of not having enough charge, or reception being poor in the region, when you most need them. The direct land line does not have these possible problems and on installation would have the emergency contact numbers posted at source, should they be required.

Many companies instigate such precautionary measures. Masters and Chief Officers of ships should insist on effective communications and not be dissuaded from instigating what they know to be right just to save money. No connection of a direct landline must be considered as false economy.

The exposed height of a vessel in dry dock could also present problems for firefighters. Clearly, stability is not an issue while the dock remains dry. However, to bring firefighters to the face of a fire on board necessitates safe access to the ship while in the dock. Two gangway access points should be a standard feature of vessels in dock. On boarding, the brigade officer should be presented with the fire arrangement of the vessel and relevant details related to the fire.

Dock personnel are obviously aware of the fire risk in dry dock and to this end regular gas freeing takes place at frequent intervals, especially on tanker vessels. Senior officers should be aware of the potential risks associated with tanks and spaces which have not been regularly gas freed.

Safety meetings between dry dock and ship's personnel are now expected to take place each morning, before the working day. Such meetings are meant to disclose any potential fire risk as part of their agenda. They would also generate dated and timed gas-free certificates for the vessel being worked.

Passenger Ship Fires

A passenger ship is defined as a ship which carries more than 12 passengers. Passenger vessels are constructed to high fire specification. Accommodation spaces are protected by A-60 category bulkheads with deckhead ceilings and linings being constructed of non-combustible materials.

Such vessels are also fitted with smoke detectors and a fire detection and alarm system. They will also operate with a fire patrol system and a continuously manned central control station. Full details of fire protection structures and systems can be found in Chapter 2 of SOLAS, parts A, B, C, D, E, F and G.

Passenger vessels have the potential for incurring high casualty rates in the event of an outbreak of fire on board. Large numbers of persons finding themselves in the confines of a ship at sea can also expect to experience signs of panic where fire is present. Crew members, although theoretically trained in crowd control, may be at a loss as to what to do in a real-time emergency.

This is not stated to put crew members down, but few will have been involved in an actual emergency. Training and drills are all-important, but when the reality of smoke effects and flame are on hand, the level of tension can be expected to rise. Subsequent levels of panic could be a side-effect, especially among passengers with a limited marine background.

Fire fighting procedures on-board passenger vessels have endemic problems. Accommodation alleyways are long and with limited space for firefighters to manoeuvre. Public rooms, like theatres, bars and restaurants generally have large, open spaces but can be cluttered by furniture and fittings. The larger vessels with multi-tier decks can limit accessibility to firefighters bearing awkward equipment.

Figure 4.8 Multi-deck passenger vessel the *Costa Atlantica* seen from the stern with high freeboard and high superstructure, which could place limitations on firefighters.

Fire Fighting Apparatus

Figure 4.9 A typical internal fire station found inside the alleyway of a passenger cabin accommodation area. Hose connected directly to hydrant, wheel key for valve operation and fire extinguisher. Above, a testing station for the local sprinkler system section.

Fire on Board a Container Vessel

The modern-day container vessels carry many thousands of container units stuffed with all kinds of trade goods, including cargoes which are classified under the IMDG code. Such goods have to be declared and the nature of the commodities are then stowed and secured appropriately, taking into account any segregation or open-deck stowage requirements. The possibility of a container within the stack which is packed with flammable goods is a reality. Provided no ignition source is in close proximity, there should be no reason for undue concern.

However, in the event of an incident, possibly with another container in the same stowage pattern, fire cannot be ruled out completely. If and when it occurs and the fire alarm system is activated, crew members could be faced with the difficult task of firefighting in and around the container unit.

Figure 4.10 The container vessel *Independent Voyager* seen under the gantry cranes at Liverpool, England. The above-deck stowage is nine units across the beam of the vessel, with an on-deck stow of four high. The container cells below decks will carry more than the upper deck total. Units are tightly packed and would present a difficult surface to permit firefighters access to inner stowed units. The upper surface of the container stack will experience approximately double the wind force that would be encountered at the main deck level in open sea conditions.

Ship's Masters would invariably make full use of the wind in the event of an on-deck container fire or toxic leakage, handling the ship to allow flames or toxics to be blown clear downwind of the vessel. A container unit on fire below decks may lead to CO_2 flooding of the hold in the event that conventional firefighting methods are not practicable.

Accessibility to inner stowed units is rarely possible and tackling a blaze with conventional firefighting equipment, while in B/A gear, in such a confined space would probably be unsustainable. Additional hydrant connections with added hose lengths is a normal feature of the lifesaving appliances fitted to modern container vessels. Such additions provide a greater ability to manage boundary cooling over a wider area.

Container ships are built with specific design features, inclusive of underdeck passage ways, cell guides for slotting container units in columns and, in some cases, no main uppermost deck. Such features will invariably influence the way fires are fought. Container vessels with no upper deck would not have the availability of total flood CO_2, but could pump vast volumes of water via the ship's bilge systems.

Virtually all unit loading vessels are structured to accommodate container units on deck in a deck cargo stowage pattern. By the very nature of hazardous cargoes it is usual to expect containers stowed on deck to be those which feature potentially dangerous cargoes, like gaseous tank containers. Any leakage/spillage of toxics or gaseous elements can then be easily vented to the atmosphere, without the added risk of accumulating inside an enclosed compartment. Firefighting or cooling of such cargo parcels, if likely to ignite, can be difficult to achieve because of the height of stowed units. Ships' equipment will now frequently include water spray lances to extend coverage to awkward and inaccessible positions.

The nature of a container's cargo dictates that container vessels cannot be allowed to roll excessively. The majority are subsequently fitted with a stabilisation system of either flume tanks or fins or a combination of both. Nor are they allowed to list when loading and discharging because the cell guides would be off line. This would not allow unit alignment and would bring loading/discharge to a halt. They also have the inherent dangers from parametric rolling and high deck stowage. This could be accentuated when in cold climates, where the upper container stack is vulnerable to ice accretion.

A fire on board might, then, seem as not the worst event in the world, but one which has to be dealt with. A course change to a 'port of refuge', where the containers can be discharged to get at a possible deep-seated fire, might also seem a prudent alternative for the ship's Master as a means of remedy. The ship's position and the geography would clearly influence any decision to deviate from the scheduled passage plan.

Fire on Board Tanker Vessels

The inherent dangers of a fire on board any vessel are considered a major hazard, but fire on board a tanker is a catastrophic situation. Any source of heat or naked flame in the vicinity of flammable liquids is totally unacceptable. The atmosphere around oil cargoes and bunker fuels is contaminated with oil vapours which are liable to ignite even in contact with warm surfaces.

In the event that the vapour ignites and fuel starts to burn, (1) it is a very hot fire and, (2) it will give off high volumes of dense toxic smoke. Gas pockets near the surface of liquids may be liable to explosive reactions, posing an additional hazard to firefighters.

Such fires are usually tackled by means of a fixed deck foam fire extinguishing system which complies with the 'Fire Safety Systems Code'. Tanker vessels of 20,000 dwt and over may be fitted with an alternative in lieu, if acceptable to the Administration, but any alternative must comply with SOLAS (ChII-2, Reg.10, 8.1.2), being capable of combating fires in ruptured tanks.

A varied number of protection systems are commercially available to high-risk spaces like cargo pump rooms and machinery spaces. Such spaces may be fitted with alternative systems employing carbon dioxide, or a fixed-pressure, water-spray, fire extinguishing system which complies with the Fire Safety Systems Code.

An anti-fire culture prevails strongly among tanker crews for the obvious reasons of self-preservation. This is not to say that complacency cannot creep in and new recruits to the industry need to be brought up with a constant awareness of the associated dangers.

Figure 4.11 3 man fire team tackle an oil fire with foam applicator.

Figure 4.12 Ultra large crude carrier the *Jahre Viking* was the largest mobile, manmade structure in the world, when built. It has now been converted from tanker operations to function as a floating oil storage unit of 564,000 dwt.

Case Incidents

The Malaysian-owned chemical tanker *Bunga Alpinia*, of 38,000 dwt, built in 2010, was believed to have been struck by lightning in August 2012 while loading methanol at the port of Labuan, in Malaysia. A series of explosions was reported to have occurred, followed by a blaze down the virtual length of the ship.

More than 80 firefighters spent several days in fighting the fire to prevent it spreading to the Petronas Chemical Terminal. At the time of writing, an enquiry is ongoing and the lightning strike has to date not been confirmed as the initial cause of the incident, although a storm was being experienced at the time.

Design for Fire Safety

On the basis that avoidance of fire is better than the cure, tanker vessels are now constructed incorporating stringent fire prevention aspects or the capability to contain any fire outbreak. The ship design is usually structured into hazardous and non-hazardous areas. These specific areas are to prevent ignition sources, while at the same time detecting combustible gases and effecting a warning system in the event of fire.

Hazardous areas are considered as not only the cargo deck, but also cofferdams, chemical stores, pump rooms, machinery spaces, paint rooms, etc. They are protected by A-60 bulkheads or decks, providing 60 minutes of protection from smoke and flame.

In comparison, the accommodation structure corridors, stairways and service spaces are classed as non-hazardous areas of considered low risk. Normally these are protected by B-0 class panels, providing 30 minutes of protection from smoke or flames.

Doors between A-60 bulkheads must be A-class doors fitted with self-closing devices. Accommodation doors are B-class doors fitted with magnetic closing devices and linked to the fire alarm system.

High-risk areas that are expected to be protected by spray nozzles include boiler fronts, above bilges and tank tops where oil fuel is likely to spread, oil fuel units like purifiers and clarifiers, hot fuel pipes near exhaust systems or similar heated surfaces.

Additionally, it should be noted that machinery spaces of Category 'A' in passenger vessels carrying more than 36 passengers must be provide at least two suitable water-fog applicators.

5
Abandonment

Introduction

The ship itself has always been considered as the best 'lifeboat'. It would only be abandoned where the vessel becomes unsustainable. The ship provides all life support systems inclusive of shelter, sustenance and communications. Its very character is meant to preserve life, and persons on board any vessel would be reluctant to give up this level of security.

However, where a vessel experiences an incident of major proportions, say from collision or grounding, where the stability and wellbeing of the ship becomes a concern, abandonment may be the only viable option. The second line of defence of any ship is by means of the survival craft. The regulations now require that all vessels carry sufficient survival capacity for all persons on board. Should the parent vessel sustain catastrophic damage of *Titanic* proportions, this secondary lifeline would take over when in open waters.

NB. The cruise ship *Oasis of the Seas* carries over 6,000 passengers and over 2,200 crew members. Operated by Royal Caribbean, she is one of the largest vessels afloat. The vessel is fitted with 18 lifeboats, each with a capacity of 370 persons, together with additional liferaft capacity.

Where the vessel is in shoals the lifeboats and liferafts may still be usefully employed, but it should be realised that alternatives may present themselves as an option. Landing personnel direct to a land mass may be a more positive safety action, e.g. the *Sally Albatross* landed passengers direct to an ice surface by means of an MES when grounded in the Baltic Sea in winter.

There are obvious concerns for abandonment with any vessel, but none more so than with the large passenger vessels. Associated problems arise with large numbers of personnel who find themselves in an unfamiliar marine environment. The human elements of anxiety, helplessness, fear or panic can be expected to rise to the surface, among crew members as well as passengers, although a well-trained crew, who have experience of crowd control and who can demonstrate assertiveness would expect to fair better than persons who are without such training.

In any major incident the parameters will differ considerably. The weather will probably be a major factor in success or failure of any abandonment. The language of commonality will influence clinical movement of large numbers of personnel. Knowledge and experience of equipment will reduce accidents and operational problems. Prudent use of communications may in itself prove a saving grace where a major incident can be handled without injury or loss of life.

Loss of the Ship

The hostile environment of the sea is well known and an accepted fact. Well-built, powerful ships are constructed to overcome the harsh elements. However, occasionally the sea will dominate and a vessel will succumb and may sink. Any loss of any vessel is one too many, but it happens and we have grown to accept that a ship may sink. Nothing is unsinkable. How the vessel actually goes down will be variable. The *Titanic*, so we are told, gradually lowered itself by the head before the aft end moved into the vertical. Enough time to launch the limited number of lifeboats in an acceptable manner. But ships do not always go down in a gradual way. Many vessels during the Second World War, after being torpedoed, capsized quickly before going under the water. This sudden motion gave sailors little chance to launch any form of survival craft.

More recently, the *Herald of Free Enterprise* (1987) capsized off Zeebrugge, but didn't sink. The depth was so shallow that when she capsized the vessel lay on her side, in the shoal water. The accident was the worst to affect a British ship since 1914, and cost 193 lives of passengers and crew.

In the case of the *Herald of Free Enterprise* she was equipped with all standard lifeboats, liferafts and marine evacuation systems (MES) to each side. Once capsized on her side, the lifeboats could not be launched, MES could not be deployed and the internal alleyways became like vertical lift shafts. The only lifesaving equipment of practical use in the immediate aftermath became the throw over liferafts, lifebuoys and lifejackets worn by personnel.

We anticipate ships may sink, which in itself is a disaster which we have to accept. When a vessel sinks by the head or by the stern and she remains upright it gives seafarers a chance to launch survival craft. However, in a capsize situation, time may be extremely limited and launching devices could be rendered useless. A similar situation occurred with the recent grounding of the *Costa Concordia* (2012) passenger vessel. The vessel listed about 80° over to starboard, making the evacuation and launch of survival craft precarious. Thirty-two persons lost their lives, mostly those who were caught below decks during the period of capsize.

NB. The *Costa Concordia* was refloated and moved in October 2013 to be dismantled and scrapped. At the time of the disaster she was carrying more than 4,250 passengers.

The Aftermath of the *Herald of Free Enterprise*

Following the capsize of the *Herald of Free Enterprise* off the port of Zeebrugge, the vessel was righted and salvaged intact but damaged. It was hoped the vessel could be refurbished and sold but no buyers were found. The vessel was towed to Taiwan and scrapped in 1988.

Figure 5.1 The Townsend Thoresen ro-pax ferry *Herald of Free Enterprise* seen righted and recovered after the capsize. Initially she was towed to Zeebrugge, where the remaining bodies were removed and then towed to Flushing. Her name was changed to *Flushing Range* for her final towed voyage to Kaohsiung in Taiwan, where she was finally scrapped by March 1988.

The Loss of *Costa Concordia*, Passenger Cruise Ship

Probably one of the most prominent maritime disasters of the shipping world occurred when the passenger cruise liner *Costa Concordia* (114,137grt) struck the rocky foreshore of Isola del Giglio, a small island off the west Italian coast. The accident happened on 13 January 2012 at about 21:45 hours, in calm seas and overcast weather conditions. Thirty-two passengers and crew were to lose their lives following the ship tearing a 50-metre gash in her port side as she manoeuvred too close to the shoreline.

The damage immediately flooded the engine room, causing a loss of propulsion and power to electrical systems. The ship developed a list and started to drift back to the island, where she grounded in a partially capsized position 500 metres north of the village of Giglio Porto. The vessel came to rest on her starboard side in shallow water, carrying 3,229 passengers and 1,023 crew members.

Figure 5.2 The *Costa Concordia*, built by Fincantieri of Italy and operated by Costa Crocier. At 114,137 grt she was developed in 2004 as the largest Italian cruise ship ever conceived, at a cost of $570 million. Titan Salvage have obtained the contract to salvage the vessel at an estimated cost in excess of $600 million during 2013, making the salvage operation the most expensive. It is anticipated that the vessel will be refloated and towed to an Italian port and cut up for scrap, becoming the world's most costly salvage operation.

This incident was significant for several reasons, mostly because it occurred while the ship was conducting a 'sail past salute', which seemingly made this an unnecessary accident, if such a phrase is conceivable. Other areas of concern established that the order to abandon ship was not actually given until over an hour after the impact took place. This is worthy of note because international maritime law requires that passenger ships must be capable of evacuation within 30 minutes.

The ship's Master has been accused of manslaughter and been placed under house arrest. Five other members of the crew, including the ship's Chief Officer and three of Costa Cruises' shore-based staff, are currently facing criminal charges.

This prominent incident happened in full view of extensive media coverage, inclusive of television, radio and the press. The accident details were released for public consumption as they occurred. However, a capsized deck where people could not stand upright can distort what can and cannot actually be done by those persons directly involved. Ship's personnel were clearly facing a potential trial by media.

At the time of writing this work, the inquiry into the *Costa Concordia* had not been made public, but the value of a safe passage plan, for any voyage, must be seen as an essential element. Deviations from an acceptable plan must be justified by Watch Officers and Masters. Passage plans, once formulated prior to the commencement of a voyage, are not meant to be written in tablets of stone. The execution stage of any plan must incorporate flexibility to accommodate prevailing conditions, but flexibility should not embrace unsafe practice. Nor is the devised plan meant for the use of a single individual. All Watch Officers and bridge team members are expected contributors to the successful execution and final outcome of any voyage plan.

It is understood that the passage plan had been updated to accommodate changes to include a 'sail past salute' to way points within half a mile of land. I believe most seafarers would consider any distance of less than two miles to be a close-quarters scenario.

Additionally, it is reported that the Master arrived on the bridge at 21:34 hours and noticed that the ship was on auto-pilot steering control. The vessel at this time was just 11 minutes from impact. The Master is believed to have ordered a change from auto-pilot to hand steering and taken the conn at 21:39 hours. The ship was 0.7 miles away from the approaching wheel over point and moving at a reported speed of 15.5 knots. It is appreciated that the speed element is a requirement of controlled steerage, but high speeds operating in confined waters will effectively reduce a person's available ahead-thinking time that much quicker.

At this time the Master is reported to have observed foam in the water ahead and ordered an alteration of course to starboard. The impact then took place at 21:45 hours. The VDR showed that the rudder angle was set at 20° to starboard ten seconds before 21:45 hours, the reported time of impact.

Author's Note

Past experience and company standing orders generally state that when navigating in close proximity to coastlines, the Master would always be in a position on the navigation bridge, taking the 'conn' well before closing any land mass. Neither would a ship's Master expect to have his vessel engaged in auto-pilot when close to a shoreline. Manual steering would also normally be expected to be engaged when entering and leaving ports and harbours or other in-close shoreline navigational situations.

Additionally, the 'Bridge Procedures Guide' stipulates navigation officers taking over the watch and control of the navigation bridge should attend the bridge with adequate time in hand. This is to allow Watch Officers to acclimatise their eyes and for operational systems to be taken into account.

If the incident of the *Costa Concordia* is examined, immediately prior to the impact, the question has to be asked, what was the Master of the ship doing minutes before the impact occurred? The Master's prime responsibility, with the ship so close to the land, must have been the safe navigation of the ship. No other subject should have been a consideration or a distraction from this prime responsibility.

It is always regrettable when loss of life is involved and the report on this incident is of great concern to mariners and potential passengers alike. It would be unsafe, at this stage, to offer any opinion without detailed analysis of the findings of the report.

Once all the facts are known it is hoped that the industry will learn and gain any benefits for the prevention of such incidents in the future. Speculation on the basis of hindsight is never helpful and the official report should clarify the what, when and why.

Figure 5.3 Salvage teams secure side buoyancy/ballast tanks to the port side of the stricken cruise liner prior to attempts at righting of the hull and bringing the vessel to the upright.

Findings from the Official Italian Report on the *Costa Concordia* Disaster

Some 17 months after the incident of the *Costa Concordia*, where 32 persons lost their lives, the official Italian report found that the human element, not just one single individual, was the root cause. However, it also found that the Master's 'unconventional behaviour' was the main cause. Blame was additionally placed on the bridge team for not attending the ship's position and on the lack of any challenge to the Master's decisions.

The report also highlighted failures of the Designated Person Ashore (DPA) to grasp the seriousness of the incident and who ordered the ship's Master to speed up the ship's evacuation. The DPA, Mr Roberto Ferrarini, was given a sentence of two years and ten months. Two other company officials were also to receive short prison terms for pleading guilty to charges of 'multiple manslaughter'. Sentences of under two years were suspended, while longer sentences would probably be replaced by 'house arrest' or community service.

The early findings also disclosed that inappropriately scaled charts were in use at the time of the accident (scale 1/100,000 instead of 1/50,000) and suggested that the operational use of ECDIS needed additional attention. Another point of concern was that the general emergency alarm was late in being initiated and this might have slowed down the abandon-ship procedures.

An interesting aspect raised was why the bridge team did not challenge the Master's decision to change the voyage plan and subsequently warn of the impending danger. Such a statement must be interpreted as that any member of the bridge team must have the right to question any and all actions of the Master. The interpretation is that the authority of the Master is no longer seen as sacrosanct. If this is the case, it has serious consequences for junior officers finding themselves in difficult circumstances.

The realisation that the Master may not always be right moves the level of responsibility to Watch Officers and, in particular, to the shoulders of the Chief Officer. Relieving a Master of command in a court of law is one thing; relieving a Master of command on the bridge of his own ship smacks of mutiny and tends to be against every belief that seafarers have ever known. For anybody to countermand the Master's orders or assume command, the reason must be unequivocal, as with mental illness or injury to the ship's Master.

The interim reports into the *Costa Concordia* seem to offer more questions than answers, especially into the actions of the Master and crew. The navigational procedures and the future design of ships to counter flooding may well be called into question by future changes to SOLAS. Some comparison can be made between the torn hole down the side of the ship and the damage sustained by the *Titanic* when it was struck by an iceberg, causing her to sink in open, deep waters, while this recent cruise ship incident was generated by an unnecessary manoeuvre in restricted waters.

The report found that the 53-metre gash in the hull below the water line led to massive flooding, causing a rapid loss of power and key service failures. Emergency pumps situated in damaged compartments failed and became overwhelmed by the sheer volume of water inside the hull.

A subsequent paper tabled at IMO has suggested the segregation and redundancy of vital equipment which directly effect steerage, control and navigation. The paper additionally draws attention to double hull protection becoming a necessity for watertight compartments, especially from side raking damage.

Improvements to bridge management and muster lists should also be of concern and a review of safe manning principles applicable to large passenger vessels should be a consideration.

Costa Concordia, January, 2013

The salvage on the Costa Concordia caused a platform to be constructed next to the offshore port side of the stricken vessel. This platform would allow the vessel to be rolled upright once buoyancy/ballast tanks had been secured to the port side. Adding buoyancy tanks to either side will increase the water plane area and so enhance the stability of the damaged ship. The subsequent movement generated by par-buckling would then bring the vessel upright onto the constructed platform. This would allow the ship to be fitted with the additional buoyancy/ballast tanks to the starboard inshore side.

The ship is expected to be taken under tow to an Italian mainland port where she will be scrapped. The salvage operation is recorded as the most expensive marine salvage task ever, costing in excess of $800 million. It was also the largest ship to have ever been salved by par-buckling.

Figure 5.4 The Costa Concordia lies aground on the starboard side, while supporting side tanks are secured to the port side.

Reflections on the *Costa Concordia* Accident, 12 Months On

A year after the *Costa Concordia* incident, the initial findings revealed that the ship had sustained a 53-metre hole, down its port side from contact with the rocky foreshore. This tear broached three watertight compartments by way of the engine room and caused unmanageable flooding within 45 minutes of contact.

The broached compartments included machinery spaces, which included the essential pumping arrangements that under normal circumstances probably would have been able to contain the flooding element. This fact is expected to be changed in future ship design to ensure that pumps are more readily available down a greater length of the ship.

Clearly, flooding after sustained damage must be an expected outcome. Such a scenario should be containable in the twenty-first century and to this end it is expected that the SOLAS design criteria may need to be changed. Probably greater inspection will initiate increased subdivision in vulnerable areas, like propulsion spaces, power supply areas and pump rooms.

Flooding

Many emergency incidents can lead to extensive flooding, which in itself generally leads to abandonment in the worst-case scenario. Collision damage, especially below the waterline, can generate a fast ingress of sea water. The deeper the damage below the waterline, the greater the water pressure and the faster the rate of ingress. High-capacity SW service pumps will have a difficult task to hold flooding levels back, even if they can be activated. Once the space containing pumps is itself flooded, unless there is a remote starting position these pumps may not even be activated at a time when they are most needed.

A major flooding incident will render the space uninhabitable and if watertight doors are not closed immediately after the incident additional compartments will also be lost. The prudent Master should consider pre-empting the potential disaster scenario by closing watertight doors beforehand, i.e. prior to entering fog areas or restrictive coastal waters, or where reduced under-keel clearance is expected.

Once damage from a collision or grounding has led to flooding, the Master or Officer in Charge should be concerned with taking a realistic view of the loss of watertight integrity. Account must be taken of the size of the damage, those compartments already with lost buoyancy and the capacity of SW pumps that can be activated. The best that pumps can be expected to achieve, if they can be activated, is to buy additional time to allow survival craft to be made ready for immediate launch.

Flooding will possibly be better contained by increased internal design by way of greater subdivision. Additional pump capacity, together with duplication of remotely operated pumps, may improve survivability rates. Greater pump distribution throughout the ship's length may also be useful. All of these will be expensive additions to recognised construction methods. However, put against the loss of ships like the *Costa Concordia*, can we afford not to incorporate such changes in future design?

Abandonment Psychology

The very thought of abandoning a vessel at sea is alien to the norm. Yet if the need arose, individuals would tend to look beyond the actual deed of abandoning the vessel. The will to survive could be expected to kick in and individuals would do what they have to do in order to stay alive. Logical thought could be expected to override the immediate danger within a well-trained crew.

Where large numbers of passengers are present, they are generally not well trained in survival practices and as such are influenced by fear of the unknown, causing panic. They could expect to become close observers of the unusual activities of launching survival craft. They would see friends and colleagues wearing unfashionable lifejackets, engaged in anxious communications. All surrounding activities outside the boundaries of normal behaviour are likely to generate uncertainty.

The realities of bad weather or seeing injured personnel would influence what activities individuals carry out. To this end the average person would expect to follow instructions in an unknown environment, provided that persons in authority can exhibit confidence and assertiveness in managing the ongoing emergency.

Figure 5.5 Passengers assemble wearing lifejackets at a family/ passenger muster point during an abandon ship drill aboard a Class 1 cruise ship.

Family groups, if separated, will only strive to stay together and it would not be a practical option to maintain positive movement of would-be survivors unless their loved ones are close by.

Commercial vessels, with their generally experienced crews, do not have the burden of passengers within the melee of the emergency. Long-serving mariners have usually been trained up to expect the worst of any situation. They tend to have an acceptance that hazards go with the job, so to speak. Their past training will rise up and generally cause them to do what needs to be done in each and every emergency. This is not to say that the most experienced of men and women will have all the answers for every situation. But the person educated to face the emergency is that much better prepared than the passenger with, at best, a limited maritime background.

Passengers' Fear of the Unknown

The cruise industry has seen phenomenal expansion over the last two decades. Ships have become much larger and more sophisticated, but not unsinkable. Many more passenger berths are available and ship capacities of more than 7,000 persons to an individual vessel are no longer out of the ordinary. With such increased numbers there is an obvious greater scope for generating panic among groups of untrained and generally uninformed persons. Such groups having inherent problems can be expected to contain a cross-section of society, inclusive of males and females, children, the elderly and persons with disabilities.

To effect control over the multi-decks of a large Class 1 cruise liner would need a cohesive crew who know the geography of the vessel and can act as a united team with effective communications. Associated problems of high decks, angled decks if the vessel is listed and crowded exit points could be expected to cause notable levels of anxiety. Such conditions would not be helped with the exhibition of lifejackets on all persons and/or bad weather prevailing.

It must be anticipated that the atmosphere would be highly charged with fear of the unknown. In such an arena the ship's officers and crew would need to show confidence in their activities and assertiveness in their control of passenger groups.

Figure 5.6 Life support craft seen in the stowed position at the boat deck level aboard the cruise liner *Queen Mary 2*.

Passenger Movement Following an Incident

In the event of a drill or a real-time emergency being declared, large numbers of passengers would be ordered to assemble in the public spaces like lounges, cinemas and entertainment rooms. The location of these muster points is usually in close proximity to the boat deck or the embarkation decks. The problem is then to move anxious passengers towards the survival craft from the initial muster points.

The movement of large numbers – say, 400 people – at one muster station would be difficult to control so that people are not lost between the muster point and the embarkation point. It is suggested that groups of 25 at a time could be effectively controlled. This number is also compatible with the majority of survival craft capacities, e.g. davit-launched liferafts usually launch with 25 people at full capacity. Lifeboats with 50-, 75- and 100-person capacities are common to passenger vessels.

Movement can be achieved by people forming a 'crocodile', putting the hand on the shoulder of the person in front. At the lead position a crew member leads the crocodile line towards the embarkation deck. The tail end of the line is taken up by another crew member.

Crew personnel who expect to be involved with passengers must now undergo crowd control training. Such training includes the lead and tail positions of crocodile lines as a method of delivering manageable group numbers towards survival craft.

Figure 5.7 Passengers wearing lifejackets participate in group movement by means of the crocodile line system. Once delivered to the survival craft station, names would be taken and people boarded prior to launch.

Figure 5.8 A partially enclosed lifeboat, seen stowed between davits aboard a ro-pax ferry operating out of the United Kingdom. The self-launching control wire is clearly visible at the aft end extension arm. Passenger access is via the stair platform once the roll-down enclosure cover is cleared.

Passenger Behaviour

For Muster and Abandonment

SOLAS regulations require passenger vessels to be able to launch survival craft within 30 minutes of the signal to abandon ship. This is clearly not the same as having the capability of evacuating all persons on board in that short time space. Regulations stipulate that essential systems must be maintained in operational condition for at least a three-hour period to ensure an orderly evacuation of personnel (MSC Circular 1214$_{16}$). Such systems include internal and external communications, emergency power and pumping systems for bilges and fire mains.

A ship's lifeboats are now no longer restricted to a maximum of 150 persons and IMO has accepted that novel life-saving appliances have been and will continue to be developed. The larger generation of vessel, already carrying many thousands of passengers, need and require modern systems like the RFD 'Marine Arc' vertical evacuation chutes. Higher-capacity liferafts and lifeboats are needed to accommodate greater numbers of persons who may be involved in a major evacuation incident.

Oasis of the Seas carries nine lifeboats on each side, each with 370-person capacity, in a fixed davit system with the boats permanently positioned overside. The davit design (LS45-Schat Harding) removes the need to swing davits outboard to achieve launch.

Mustering Passenger Numbers

There are many potential problems associated with a mass assembly of large numbers of persons. Many large passenger ships muster their passengers in respective 'zones' of the vessel. These zones are generally linked to the stateroom/cabin areas and employ a public space like a cinema or dining area to accommodate 500–600 people to a muster position. These muster stations are rarely on the 'boat deck', but in close proximity to enable access to boat boarding positions. This geography allows an element of control to prevail in assembling persons and their movements from the muster points to boat seating arrangements.

Communications at the muster points may be difficult in terms of language and pronunciation of instructions. Such an example could be seen with the calling of 500 names, taking well over an hour in what may be a limited time envelope. Commercial companies have developed a 'passenger accountability system' by using scanning devices on barcoded passenger lifejackets, identity cards or individual cruise boarding cards. The scanners are linked directly to the passenger manifest computer network.

Systems have improved to deal with greater numbers of people, but no system is infallible. Circumstances will dictate levels of survivability; the only consideration is to learn from each incident and continue research and development into self-improvement.

Incident Report

In May 2006 the *M.V. Calypso*, a passenger vessel carrying 708 passengers, experienced an engine room fire in the starboard engine. The vessel was en route from Tilbury to Guernsey in the Channel Islands and was in a position approximately 16 miles off Eastbourne in the English Channel when the fire evolved.

The Master ordered the passengers into the ship's lifeboats but did not launch the boats, and the crew tackled the fire. This action was clearly a precautionary one, in that the fire could be contained and extinguished, which was eventually achieved.

The boats did not launch and with hindsight the action of moving passengers to the boats ready for disembarkation provided real-time experience for seafarers in the emergency movement of civilian personnel. The ship suffered extensive damage from this incident but no casualties.

Figure 5.9 Passenger vessel *M.V. Calypso* seen moored alongside a berth the height of the freeboard provides some idea of the distance that survival craft would have to be launched through in order to reach the surface. Such a launch height would pose problems for persons with limited marine experience in good weather; launching in an emergency in bad weather could be an anxious time, even for experienced personnel.

Exposure to Risk

Individuals who enter the maritime environment deliberately and usually voluntarily expose themselves to a level of risk. For the most part this is a tolerable risk, but if they find themselves in an abandonment situation, exposure to immersion in sea water becomes a much greater possibility.

There are known to be four phases associated with immersion, leading to the death of the individual from acute hypothermia. These phases are:

1   initial immersion
2   short-term immersion up to 15 minutes
3   long-term immersion of 30 minutes or more
4   post-immersion, leading to collapse and death.

When ordering any ship to be abandoned, persons in authority should be clearly aware that individuals are at a much higher level of risk from the effects of immersion. It is therefore essential that any evacuation methods from a parent vessel strive to keep persons dry and avoid contact with the water.

The temporary sanctuary of enclosed lifeboats and liferafts cannot be underestimated where the loss of the parent vessel has occurred. Where possible these should be launched in such a manner as to keep the occupants dry. The increased threat level to persons who have to enter the water prior to boarding a survival craft is considerable.

Full use of immersion suits or other similar waterproofs worn over layers of warm clothing will effectively trap layers of warm air around the individual. These layers will increase protective levels against the onset of hypothermia and extend life expectancy to persons who may have been forced to enter the water.

The employment of current life-saving appliances tends to be adequate for a standard sinking situation, if there is ever such a thing. However, when a capsize is involved, large numbers of personnel, as with passenger/ferry vessels, will be denied the luxury of a dry, conventional launch in a survival craft. The use of hydrostatic release units (HRUs) fitted to survival craft may prove themselves an ideal format to allow survival craft to gain a position of comparative safety. This would mean personnel must expect to enter the water to reach these in order to subsequently obtain any benefit. Such action would increase the likelihood of immersion and heat loss occurring to the body.

The temperature of the sea water in the position of an incident will be a critical factor affecting the casualty list. Where the water is very cold – like Alaska, Antarctica and the far North Atlantic – the low temperatures are detrimental to prolonged survival. Many persons entering the water would experience 'cold shock' and not survive the first phase. Persons adequately clothed and protected and wearing a lifejacket should be able to sustain themselves for an interim period of time, but not indefinitely.

Persons passing through a period of long-term immersion could expect to experience all the symptoms of acute hypothermia to the point of unconsciousness. At this stage the person could easily be mistaken for dead.

Evacuation by Free Fall Lifeboat

The use of 'free fall lifeboats' has become widely used aboard new commercial tonnage. It is a system designed to accommodate the total complement of the vessel and as such would not be appropriate in its present form for passenger vessels.

Following an order to abandon, all crew members can enter the craft from the stern and it can be launched from an internal position. Personnel are strapped in with a full-body harness to take account of the gravitational impact at the surface.

Many lifeboat systems are launched and recovered by wire falls. Historically, the on-/off-load hooking arrangements for operations with falls have caused numerous accidents. This necessitated the introduction of 'fall preventative devices' to be introduced and used at the Master's discretion. The free fall system is not affected by such changes. Current modifications are taking place on on-/off-load systems and manufacturers are taking steps to ensure safer launch systems and avoid boats going into free fall from either end.

Figure 5.10 Boarding arrangement and support structure for a totally enclosed free fall lifeboat, situated for over the stern launch from the vessel.

Figure 5.11 Free fall lifeboat seen in the stowed position, aft, ready for free fall launch.

Totally Enclosed Lifeboats

The majority of lifeboats at sea today are of the totally enclosed or partially enclosed type. Some older tonnage trading in warm climates are still allowed to operate with open boats but these are gradually becoming fewer in number.

The obvious benefit to the use of enclosed boats is increased protection from cold weather conditions. Additionally, the fully enclosed types can be sealed against toxic gases and give a level of fire protection that is denied with liferafts. The enclosed system incorporates a water sprinkler system as well as a compressed air bottle to provide such protection for the occupants.

Standard equipment is inclusive of a 24-hour fuel supply for the engine. A standard lifeboat can be a designated rescue boat, provided it meets the standards and equipment as required for rescue boats.

Maintenance Programme for Life Saving Appliances

Weekly	Boat engines must be operated and boats moved from chocks.
Monthly	Boats' equipment must be checked and inspected and seen to be in good order.
Quarterly	Lifeboats must be lowered to the surface and manoeuvred in the water.
Annually	12-monthly inspection under the Safety Equipment Certificate (conducted by the Marine Authority).
Five-year interval	Lifeboat falls must be renewed and the launching system must undergo a load test to 110 per cent of the safe working load (SWL).

Crews are expected to exercise in abandon ship and fire drills weekly on passenger ships. Cargo ships are expected to conduct at least one abandon ship and one fire drill in which every crew member must participate on a monthly basis.

Drills provide confidence and reassurance to both passengers and crew where they are seen to be conducted in a positive manner, as well as useful prior knowledge in the event of a real-time incident occurring.

Evacuation by Inflatable Liferaft

Liferafts are generally not gas tight and could be affected by flames at surface level, as from burning oil. They are difficult to board by

Figure 5.12 Inflatable liferaft seen in the inflated condition, tethered in sea conditions.

persons in the water and would be subject to damage by persons trying to jump directly into or onto the canopy, from height, or from an embarkation ladder.

They represent a second line of defence alongside lifeboats, but they remain vulnerable in open sea conditions. Occupants would benefit from a level of basic marine survival training, which passengers in the main would not have, though a set of instructions on the use of the raft and its equipment are included as standard issue.

An undamaged liferaft will have enough buoyancy for a 100 per cent capacity overload; however, floor space would be limited and rations would not sustain total double occupancy for any length of time. Standard inflatables use CO_2 mixed with a small amount of nitrogen to cause two main buoyancy chambers to inflate. These are susceptible to puncture damage from sharp-edged floating debris. One of the duties of any lookouts posted would be to be alert for flotsam that could pose such a hazard.

It would be expected that liferafts, being without any form of mechanical movement, would become the target of rescue boats in a marshalling operation, assuming the rescue boat itself was able to clear the immediate vicinity of the abandonment.

Figure 5.13 Inflatable liferaft seen in cradle stowage at the ship's side. The canister is secured by straps and fitted with a disposable hydrostatic release system.

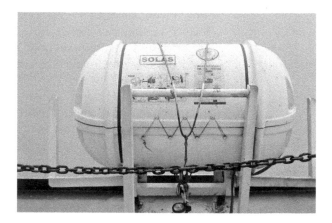

Figure 5.14 Open-deck davit-launched liferaft station seen on the foredeck of a passenger vessel. The liferafts are stowed against the bridge front (not in image).

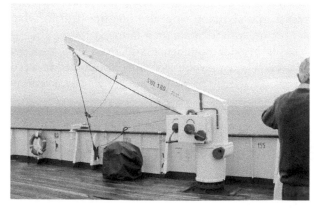

Evacuation by Davit-Launched Liferaft

This launch system for liferafts is now more prolific than ever. The reasons for this are that the number of passenger ships has increased and davit-launched systems are systematic to these vessels. Also, many more vessels are being fitted with free fall lifeboats, and davit-launched systems tend to accompany ships so fitted because of the inherent recovery system of the boat. Such a recovery system provides a dual role as a launching method for liferafts.

The main and obvious advantage of the davit-launched system is that persons can board dry and stay dry throughout the period of occupancy. Also, injured persons can be lifted directly into the raft without having to climb down embarkation ladders to reach waterborne survival craft.

As with any other liferaft, a degree of prior knowledge in managing and handling would be a distinct advantage once it has been successfully launched. The main disadvantage is that the procedure to carry out the launch is more detailed than launching the throw-over inflatable. Personnel must be loaded in a stable manner and the person in charge of the raft must activate the release when closing the surface.

The system is such that multiple liferafts can be launched from the single-arm davit reasonably quickly. The davit is turned out to a launch position once only, yet can systematically launch ten or twelve liferafts from the one station. Liferafts are usually stacked or racked in a position close to the davit.

Figure 5.15 Example davit-launched liferaft station at embarkation deck level aboard a large Class 1 passenger vessel. The davit is fitted with a self-launching system. Disembarkation by davit-launched liferaft.

Figure 5.16 A disembarkation station aboard a passenger ferry, *Pride of Bruges*, seen with two davit systems against the ships rails. Eight canister liferafts are seen against the bulkhead. The station is equipped with embarkation lights and a loudspeaker linked to the public address (P/A) system. Embarkation ladders are seen at each end of the station stowed behind the davit arms. Deck space is limited and this would force a control system on the number of passengers allowed into the area during a practical launch procedure. Crew members would need manoeuvring space to manhandle raft canisters from racks to fall hooks. Crowded deck space would restrict raft launching preparations to be conducted in a manner conducive to the situation.

The Role of Rescue Boats in Abandonment

A rescue boat has two essential functions: (1) to be able to recover persons from the water; and (2) to marshal survival craft together.

Rescue boats, either side of passenger vessels, can play a beneficial role in a sinking incident or a capsize situation, provided they can be cleared away in sufficient time. Rescue boats must be geared to recover persons from the water quickly and deliver them to a safe haven or to a standard lifeboat. Designated rescue boats carried by commercial vessels can be one of the conventional lifeboats carried by the vessel, provided that it is equipped and can perform to the regulation standards for rescue boats.

All rescue craft so carried must be fitted with means of recovery which may be in the form of a boarding ladder or recovery net. Ideally, persons should be drawn into the boat from the water in a horizontal posture, rather than being lifted out in a vertical motion. Such a recovery style reduces the hydrostatic squeeze on the person's body and reduces the risk of cardiac arrest taking place.

The duty of the coxswain of a rescue boat is to marshal survival craft together, preferably upwind. This is achieved by use of the 50-metre towing line, which is also part of rescue boat standard equipment. Liferafts have limited mobility from use of the paddles and may not be able to manoeuvre fast enough to clear inherent dangers such as high rigging making contact, should the parent vessel roll as she goes down.

Figure 5.17 *Queen Mary 2* lies port side in Southampton. Partially enclosed lifeboats seen stowed down the ship's starboard side at the boat dock position.

Passenger vessels are provided with a rescue boat on either side of the parent vessel, each fitted with a fixed means of communication. If a large passenger vessel was involved in an incident, coordination between rescue boats would become essential in collating survival craft and personnel. Accounting for large numbers of persons over a large area of open sea would not be seen as an easy task to fulfil.

Rescue Boat Operations

Launching any boat at sea must be considered a hazardous task and would normally be carried out once the parent vessel can generate a 'lee' the boat could be launched into. Any shelter from the wind would generally provide a safer environment for launching the craft from a single point bridle arrangement. Unfortunately, a generated area of lee cannot change the swell conditions and some movement of the boat at surface level must be anticipated.

Where the parent vessel is involved in an incident it should be realised that the Master may not always be able to manoeuvre the vessel to generate a lee; such examples are when power or steering is lost or when the vessel is aground. Ships' Masters would in any event expect to brief coxswains of boats prior to launching their craft in any seagoing condition.

One of the benefits of having a rescue boat on either side is flexibility as to which boat to actually launch, obviously launching the boat on the leeward side where possible. The activity of the boat, once in the water, would expect to be one of recovery of persons on its windward side, in a horizontal posture if at all possible.

Once on board the rescue craft, casualties should have wet clothing removed and replaced by dry clothes before being placed in 'thermal protective aid(s)' (TPAs) as soon as practical.

> **NB.** Although TPAs are provided as standard equipment in rescue boats, no dry replacement clothes are included as a standard. A dry overall could be beneficial.

The regulations now expect life boats (as designated rescue boats) to be launched from a parent vessel which is underway at a speed up to five knots. Such an operation must be one that deploys the boat's painter as forward as possible. The rudder action would cause the boat to sheer away from the ship's side to clear. Single point release bridle systems as fitted with fast rescue craft use the increased speed of the rescue craft to clear the immediate launch position.

Example Liferaft Operations

Figure 5.18 Military personnel under night training with the single seat liferaft. These rafts are manufactured by RFD for specific use by non-commercial aircraft pilots/navigators. They measure 1570 mm long by 980 mm wide, and have a height of 820 mm.

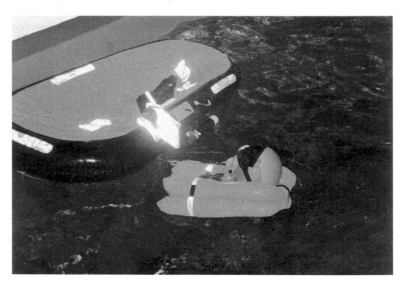

Evacuation by Means of Marine Evacuation Systems

The marine evacuation system (MES) is defined as an appliance for the rapid transfer of persons from the embarkation deck of a ship to a floating survival craft.

The first use of evacuation slides was by the *Sally Albatross*, a passenger vessel which ran aground in the Baltic Sea in the winter ice conditions of 1994. Approximately 1,500 people were relieved from the ship in about 90 minutes via the MES slides. They accommodate large numbers sliding down to a floating, inflated platform from which the survivors are moved into large-capacity liferafts.

These systems are serviced at 12-monthly intervals (the Administration can extend this to 17 months) and deployed on a rotational basis so that they are activated at least once every six years.

Figure 5.19 Marine evacuation system, seen deployed in calm and sheltered waters of an enclosed harbour. They are manufactured in single or double slide format with a raft base landing platform at the surface. Multiple high-capacity liferafts are inflated close to the platform to permit transfer of personnel from platform to raft. Vertical zig-zag chutes are now a similar alternative to dispersing large numbers of personnel from passenger vessels. Movement of persons down slides or chutes must be safely monitored to effect an orderly disembarkation.

Shipboard Emergency Drills

Both the shipping and aviation transport industries have always conducted emergency drills. The essence of a drill is to make personnel familiar with what to do in an emergency situation. They should also make people familiar with respective equipment which may not be in everyday use. Both military and commercial navies have conducted abandon-ship drills, fire drills and damage control exercises since time immemorial. They have become commonplace aboard virtually every ship and vessel afloat.

Vessels that tend to have a contracted crew with infrequent change overs for replacements can expect their personnel to become very familiar with both emergency equipment and procedures. This is especially so if 'job sharing' is a feature of shipboard policy when exercising in emergency drills. Such practice would seem to eliminate the possibility of complacency creeping in and at the same time provide greater individual experience.

Drills are considered as in-house training for crews and build passenger confidence. However, the reality is that few passengers are ever involved in 'hands on' situations and gain little from a drill other than familiarity with donning lifejackets and location awareness of survival craft. The dominating problem in many cases where passengers are involved is that a few trained crew members are trying to control large numbers of persons in a discipline that they have little or no experience of. A shepherd and his flock of sheep come to mind.

This is not meant to be a derogatory phrase, but one that reflects large passenger numbers, which generally can be expected to know little of seamanship and probably less about marine survival practices. In the main, 'common sense' will come to the fore and in serious situations the 'will to survive' could be expected to become dominant.

Evacuation by Helicopter

The use of helicopters in the form of air ambulances is well known. They are frequently used in the maritime environment to evacuate individual or small numbers of casualties. The reason for limited action is its payload capacity and the restrictive amount of fuel a helicopter can carry. If the operational range is realised for a Sea King rotary winged aircraft as being a maximum of 250 miles one way, the limitations of helicopter operations can be understood.

This is not to say that many aircraft could not be sequenced in order to carry out a mass evacuation. But adequate fuel and refuelling facilities would need to be readily available for increased numbers of aircraft. Also, large numbers of aircraft in a limited airspace must expect to generate air traffic control problems.

Some helicopters have long-range capability, especially the CH53 (Jolly Green Giants), because of their inflight refuelling capability. But the average offshore aircraft is extremely limited on operational range, unless configured with long-range additional fuel tanks. Bear in mind that the addition of fuel reduces the possible payload of the transport.

Actual recovery of casualties can be by hoist operation from a hover position or by a controlled land-on, assuming a horizontal deck space is available. Should the ship's deck be angled due to list or angle of heel, it is unlikely that a land-on scenario would be viable. Hoist operations are a practical option for a single lift, but time consuming where several casualties are to be recovered. Even if a Hi-Line operation is employed, a land-on action could take a maximum payload in the shortest possible time. As such, it must be considered as the most effective means of helicopter recovery.

Where helicopters can be brought in, the number of stretcher cases vs. walking wounded will restrict the overall payload. Some aircraft are configured for MEDIVAC, but these are limited and designed to carry medical staff plus casualties.

Helicopter engagement for a routine operation is always of concern; helicopter engagement during an emergency has the additional problems of possible disorder, fire or similar associated hazards on reception. There may or may not be a designated landing officer and safety standards that exist for routine operations may not exist. Communications with a single heli-operation generally do not pose a problem, but with the use of multiple aircraft the possibility of confusion with surface activity could be influential.

Coordination at ground and sea surface stations would be considered essential for a safe evacuation of large numbers by aircraft. For example, a passenger vessel with an operational landing area with an orderly dispatcher, controlling a flow of would-be survivors/casualties. Such activity would require a coordinating/monitoring station ashore to ensure refuelling and scheduling to and from a safe haven, with an awareness of additional activity from non-affected air traffic.

Figure 5.20 A Dauphin 2 SA365N helicopter under operations with Bond Helicopters engaged in offshore ferry operations in the Morecombe Bay area of north-west England. The hoist winch arrangement can be seen clearly above the access on the starboard side of the aircraft. The helicopter is manufactured by Aerospatiale (France) and incorporates 'fenestron technology' with the integral tail rotor. As with any helicopter engaged offshore, it has twin engines and must be fitted with flotation gear. It is a general-purpose helicopter but can be configured for military use and as a MEDIVAC alternative.

Helicopter Operations

Helicopter operations are conducted worldwide around coastal regions, but are generally restricted by payload, operational range of aircraft and also the capability of the winching system. Generally, a maximum SWL of the aircraft winch is reached with two people at the same time. Such a maximum is restrictive with large numbers of casualties. Both military and civilian aircraft tend to operate over the marine environment, frequently dedicated to SAR activities.

Figure 5.21 An RAF Wessex '5' engages in a stretcher casualty recovery exercise with a totally enclosed lifeboat. Deck space is restricted for what is termed wet deck landings.

Figure 5.22 The US Coastguard employs the Sikorsky, HH-60J 'Jayhawk' medium-range recovery helicopter (300 nm offshore) extensively around the American coastline for SAR drug interdiction and similar missions. It has capabilities to operate in high-wind situations up to 63 knots and interact from a ship's platform. It operates with a crew of four and has a lift capability of at least six persons. With good endurance to remain on station for up to 1.5 hours, this has become an extremely versatile aircraft to have around a major incident. Incident on-scene endurance is about 45 minutes.

Helicopter/Shipboard Operations

Virtually all planned helicopter engagements at sea are pre-planned via a checklist once it is determined whether the aircraft will engage in a hoist or a land-on operation. With many ships now custom designed with a defined helicopter operation site, decisions on where to engage the aircraft are frequently already established.

Assuming weather conditions permit, course alteration to favourable weather is possible. A typical checklist for engagement would expect to contain any or all of the following:

1 Define the Helicopter Landing Officer (HLO) and/or hook handler.
2 Brief bridge team personnel with HLO and deck reception team.
3 Clear the operational site of all obstructions and loose objects.
4 Display windsock or flags.
5 Display navigation signals for launching/recovering aircraft.
6 Turn out rescue boat ready for launch.
7 Establish communications between ship/aircraft as soon as possible (channel/frequency).
8 Communicate ship's name, recognition data, position, ship's course and speed and local weather conditions with sea state.
9 Stand-by to transmit homing signal.
10 Post lookouts and establish manual steering.
11 Prepare personnel for transfer with documentation and/or medication details.
12 Place main ship's engines on stand-by for period of engagement.
13 Ensure adequate sea-room and depth for area of engagement.
14 Trim deck lights to avoid pilot glare if night-time operation.
15 Lower deck rigging, aerials, cranes, halyards, etc.
16 Confirm ship's heading and wind direction with aircraft pilot prior to engagement at confirmed rendezvous.
17 Close off operational area to non-essential personnel.
18 Hook handler equipped with rubber boots and gloves, plus insulated hook.

Additional for Land-on Operations:

1 Fire party on stand-by, clear of area but ready for intervention.
  • Emergency crash-box and firefighting equipment readily available. (Equipment should include mobile CO_2 with a thermal lance – helicopter design incorporates main engines at an inaccessible height without the aid of a thermal CO_2 lance being available. The fire risk with a helicopter tends to be greater when engines are restarted after being shut down.)

2 Turn down ship rails if required.

3 Ensure deck party ready to tie-down aircraft if required.

4 Avoid communications with pilot during landing.

Additional for Tanker Vessels

- Tankers should release pressure from cargo tanks about 30 minutes before engagement. Ensure tank openings are secure following any venting operations.
- Inert gas pressures in cargo tanks should be reduced.
- Gas tankers should avoid any emissions of gas or vapours to the deck area when no inert gas system is fitted.

Additional for Bulk Carriers

- All surface ventilation to dry bulk cargoes should be ceased and all hatch covers and access lids battened down.

NB. The most critical period for a land-on operation by a helicopter, where the aircraft shuts down her engines, is when she restarts her engines for take-off procedure. The risk of an aircraft engine fire at this moment is greater than normal. Fire teams should be aware that the aircraft has its own built-in internal firefighting system and the team would not be called in to assist unless the pilot specifically signals for their intervention.

All activities towards any surface-to-air operation should be recorded in the ship's log, throughout the period of approach, contact and activity concerning the aircraft.

Ships' Masters should also consider that if the weather conditions are so severe that they could not launch the rescue boat, why is the vessel engaging in helicopter operations? It might be more prudent to wait for improved weather and provide a safer environment for all concerned.

Surface-to-Air Medical Evacuation (MediVac)

Where casualties are to be evacuated for medical reasons, the following details should be considered as additional to a routine evacuation.

General Information (to be supplied to MRCC)

- Name and position of vessel evacuating casualty.
- The name of the next port of call with ETA.
- Course and speed of the vessel from evacuation point.

Patient's Condition and Symptoms

- Medical notes to be included with passport/identity papers.
- Possible diagnosis if known.
- Current condition of patient.
- Patient's pulse rate.
- Details of medical history.
- Details of medication administered to patient.
- Any medication that the casualty is allergic to, if known.
- Whether the patient is incapacitated or ambulatory.

Weather Conditions (during evacuation period)

- Wind direction and force.
- Cloud ceiling (estimated).
- Sky condition (clear, overcast, etc.).
- Precipitation in sight.
- Sea state conditions.
- Air temperature.

Stretcher Cases

Where a litter or stretcher case is designated for evacuation, the litter or stretcher should be disconnected from the hook/hoist once it is lowered to the deck. The patient should then be strapped in, face upwards, with any documents secured inside his/her clothing. The litter should then be re-secured to the hoist prior to the lift taking place.

NB. Do not secure the hoist cable in any way or at any time throughout this operation!

Helicopter Hi-Line Capabilities

Although extremely versatile, rotary winged aircraft are handicapped by obvious restrictions, especially where large numbers of survivors prevail. For example, numerous aircraft would be required to relieve large numbers of persons from a typical modern cruise liner.

Hoist operations, other than for a single lift, are time consuming and costly on aircraft fuel reserves. To maximise its mobility, a Hi-Line technique has been perfected. This operation is carried out when there are several survivors to be recovered. Effectively, a heaving line is passed to the deck of the surface craft to act as a control line to the hoist harness(es).

The aircrew man will land on the vessel first and take charge of the operation on the deck, causing two would-be survivors to be hoisted per lift into the aircraft. Multiple double hoists would take place until the total payload of the aircraft is attained.

In this time the aircraft holds station on the vessel and the movement of persons via the double harness is achieved. The Hi-Line is slacked away and then heaved in to recover the harness for the next hoist. The repeat action transports two people upward to the aircraft per hoist.

This operation maximises the use of fuel and time in order to gain the greatest effect from the recovery aircraft. It is an operation that is preferred in bad weather, or where the ship's deck landing is obstructed by extensive rigging, like with a tall-mast sailing vessel.

It is advised that not all aircraft and crews have the training or capability to perform in a Hi-Line activity. Commercial helicopters, for instance, generally operate with limited aircrew, usually pilot and co-pilot. Designated military or Coastguard authorities are more likely to conduct Hi-Line operations.

Helicopter Incident Report

In August 2013 a super-puma helicopter transporting oil-rig workers in the North Sea was involved in a fatal crash in which four persons were lost. The helicopter was carrying 16 oil-rig staff and a crew of two.

This incident was the fifth in four years to affect super-puma aircraft. Recovery teams have recovered the black box from the helicopter and it is hoped the following enquiry will reveal the causes of the accident.

This latest incident highlights that the marine environment continues to have many hazards across many disciplines.

Miscellaneous Facts (Related to an Abandonment Situation)

- The visible range of a lifejacket light on a dark night with clear atmosphere is one mile.
- The range of the light on a liferaft on a dark night with clear atmosphere is two miles.

- A hydrostatic release system activates at a depth of not more than four metres.
- Recommended search patterns where the datum is known are: (1) sector search; (2) expanding square pattern.
- An orange smoke canister from a survival craft will burn for approximately four minutes.
- A ship-based rocket line-throwing apparatus should cover at least 235 metres.
- The temperature inside a liferaft can be raised by battening down the doors and inflating the double floor.
- The compressed air inside a totally enclosed boat is good for 10 minutes.
- Lifejackets are fitted with a towing loop to allow survivors to be towed through water.
- The ship is the best lifeboat, providing all life-supporting requirements, provided it is sustainable.
- Most marine casualties succumb to hypothermia before drowning.
- Offshore helicopters must be twin engine and fitted with flotation gear.
- Helicopter pilots must have Instrument Flying Rating in order to operate at night.
- Every survival craft must carry six hand flares, four rocket parachute flares and two orange smoke floats.
- Hand flares and smoke floats are more recognisable from search aircraft.
- Aircraft can detect a SART transmission at about 40 miles.
- Survival craft are expected to be able to launch against a 20° adverse list.
- Under GMDSS requirements, ships must carry three walkie-talkie radios with lithium batteries.
- Survivors in boats or liferafts should not eat any of the rations in the first 24 hours.
- The body loses heat 26 times as fast when wet compared to dry.

6

Marine Pollution

Introduction

The transport of oil by sea remains a dominant sector of the marine industry. The majority of vessels use oil as their main fuel as a prime source of power. It is also transported in far greater quantities as bulk cargo in the tanker trade.

There are two main hazards with the transport of oil; one is the danger of fire, the other is of pollution to the environment. Both are of major concern to Masters and crews engaged in oil transport all over the marine world.

The routine operation of taking bunkers and the practice of loading or discharging oil cargoes always have the inherent risk of spillage. As such, strict procedures for the movement of oil must be followed and ships must carry a 'Ship's Oil Pollution Emergency Plan' (SOPEP), readily available to implement. Checklists for loading and discharging of oils are available on all ships engaged in the trade and can also be found in the International Oil Tanker and Terminal

Figure 6.1 The cruise ship *Silver Wind* seen with a typical oil bunker barge alongside.

Safety Guide (ISGOTT). Designated tanker vessels also carry two oil record books, one for oil cargoes the other for bunker oil movements.

The international safety management requirements for shipping are based on safer ships and cleaner seas. As such, any oil pollution to the environment is considered a major incursion and subject to punishment by fine, imprisonment or both.

It is essential that seafarers accept their responsibilities in keeping the environment clean and clear of all aspects of pollution, not only from oils, but also from garbage, sewage, contaminated ballast water and from noxious emissions into the atmosphere. MARPOL annexes take account of all aspects of marine pollution and ship's officers should be familiar with their content and the ramifications in the event of any form of pollution.

Terminology and Definitions affecting Tanker and Gas Carrier Vessels

Associated piping – The pipeline from the suction point in a cargo tank to the shore connection used for unloading the cargo; this includes all the ship's piping, pumps and filters, which are in open connection with the cargo unloading line.

Barrier/boom equipment – Inflatable oil pollution barriers that can be deployed essentially anywhere in the world. These are used extensively at oil pollution sights to contain spillages and protect sections of the environment from damage caused by uncontrolled spills. Such equipment may be carried by the ship itself, but generally only in small amounts. More extensive 'boom gear' is usually strategically placed at emergency marine storage stations at respective geographic stations around the globe.

Bulk Chemical Code – The code for the construction and equipment of ships carrying dangerous chemicals in bulk (ships must have a Certificate of Fitness for the carriage of dangerous chemicals).

Cargo area – That part of a ship which contains cargo spaces, slop tanks and pump rooms, cofferdams, ballast and void spaces adjacent to cargo tanks and also deck areas throughout the length and breadth of the part of the ship over such spaces.

Centre Tank – Any tank inboard of a longitudinal bulkhead.

Chemical tanker – A ship constructed or adapted primarily to carry a cargo of noxious liquid substances in bulk; this includes oil tankers, as defined by Annex I of MARPOL, when carrying a cargo or part cargo of noxious liquid substances in bulk (*see also* Tanker).

Clean ballast – Ballast carried in a tank which, since it was last used to carry cargo containing a substance in Category A, B, C or D, has been thoroughly cleaned and the residues resulting have been discharged and the tank emptied in accord with Annex II of MARPOL.

Cofferdam – An isolating space between two adjacent steel bulkheads or decks. This space may be a void space or a ballast space.

Combination carrier – A ship designed to carry either oil or solid cargoes in bulk.

Continuous feeding – The process whereby waste is fed into a combustion chamber without human assistance while the incinerator is in normal operating condition with the combustion chamber operative temperature between 850 °C to 1200 °C.

Critical structural areas – Locations which have been identified from calculations to require monitoring or from service history of the subject ship or from similar or sister ships to be sensitive to cracking, buckling or corrosion, which would impair the structural integrity of the ship.

Crude oil – Any liquid hydrocarbon mixture occurring naturally in the earth, whether or not treated to render it suitable for transportation. This includes: crude oil from which certain distillate fractions may have been removed; and crude oil to which certain distillate fractions may have been added.

Dedicated ship – A ship built or converted and specifically fitted and certified for the carriage of: (1) one named product; (2) a restricted number of products each in a tank or group of tanks such that each tank or group of tanks is certified for one named product only or compatible products not requiring cargo tank washing for change of cargo.

Discharge – In relation to harmful substances or effluent containing such substances means any release howsoever caused from a ship and includes any escape, disposal, spilling, leaking, pumping, emitting or emptying.

Domestic trade – A trade solely between ports or terminals within the flag state for which the ship is entitled to fly, without entering into the territorial waters of other states.

Emission – Any release of substance subject to control by Annex VI, from ships, into the atmosphere or sea.

Flammability limits – The conditions defining the state of fuel oxidant mixture at which application of an adequately strong external ignition source is only just capable of producing flammability in a given test apparatus.

Flammable products – Those identified by an 'F' in column 'F' of the table in Chapter 19 of MARPOL.

Flashpoint (of an oil) – The lowest temperature at which the oil will give off vapour in quantities that when mixed with air in certain proportions are sufficient to create an explosive gas.

Garbage – All kinds of victual, domestic and operational waste, excluding fresh fish and parts thereof, generated during the normal operation of the ship and liable to be disposed of continuously or periodically, except those substances which are defined or listed in other Annexes to the present Convention.

Gas carrier – A cargo ship constructed or adapted and used for the carriage in bulk of any liquefied gas or other products listed in the table of Chapter 19 of MARPOL.

Good condition – A coating condition with only minor spot rusting.

Harmful substance – Any substance which, if introduced into the sea, is liable to create hazards to human health, to harm living resources and marine life, damage amenities or to interfere with legitimate use of the sea, and includes any substance subject to control by the present Convention.

Hold space – The space enclosed by the ship's structure in which a cargo containment system is situated.

Holding tank – A tank used for the collection and storage of sewage.

IBC Code Certificate – An international Certificate of Fitness for the carriage of dangerous chemicals in bulk, which certifies compliance with the requirements of the IBC code.

IGC Code – The International Code for the Construction and Equipment of ships carrying liquefied gases in bulk.

Ignition point (of an oil) – The temperature to which an oil must be raised before its surface layers will ignite and continue to burn.

Incident – Any event involving the actual or probable discharge into the sea of harmful substance, or effluents containing such a substance.

Instantaneous rate of discharge of oil content – The rate of discharge of oil in litres per hour at any instant divided by the speed of the ship in knots at the same instant.

International trade – A trade which is not a domestic trade as defined above.

Liquid substances – Those substances having a vapour pressure not exceeding $2.8\,kp/cm^2$ when at a temperature of $37.8\,°C$.

Marine Pollution Control Unit (MPCU) – The investigative branch of the MCA which investigates marine pollution incidents and reports.

MARVS – The maximum allowable relief valve setting of a cargo tank.

Miscible – Means soluble with water in all proportions at wash water temperature.

NLS Certificate – An international Pollution Prevention Certificate for the Carriage of Noxious Liquid Substances in Bulk, which certifies compliance with Annex II of MARPOL.

> **NB.** An NLS Certificate, an IBC Code Certificate or BCH Code Certificate is issued on the date of completion of the relevant survey and is valid from the date of issue for a period not exceeding five years.

Noxious Liquid Substance – Any substance referred to in Appendix II of Annex II of MARPOL, or provisionally assessed under the provisions of Regulation 3(4), as falling into Category A, B, C or D.

> **NB.** Since 2004 it has been a requirement that every ship over 150 grt which is certified to carry noxious liquid substances in bulk shall carry on board a shipboard marine pollution emergency plan for noxious liquid substances.

NOx technical code – The technical code on control of emission of nitrogen oxides from marine diesel engines, adopted by the Conference, Resolution 2, as may be amended by the organisation.

Oil – Petroleum in any form, including crude oil, fuel oil, sludge oil refuse and refined products (other than petrochemicals, which are subject to the provisions of Annex II).

Oil fuel unit – The equipment used for the preparation of oil fuel for delivery to an oil-fired boiler, or equipment used for the preparation for delivery of heated oil to an internal combustion engine, and includes any oil pressure pumps, filters and heaters with oil at a pressure of not more than 1.8 bar gauge.

Oily mixture – A mixture with any oil content.

Oil tanker – A ship constructed or adapted primarily to carry oil in bulk in its cargo spaces; this includes combination carriers and any 'chemical tanker' as defined by Annex II when it is carrying a cargo or part cargo of oil in bulk.

Organisation – The Inter-Governmental Maritime Consultative Organisation or the International Maritime Organisation (IMO)

Permissible exposure limit (PEL) – An exposure limit which is published and enforced by the Occupational Safety and Health Administration (OSHA) as a legal standard. It may be either time-weighted average (TWA) exposure limit (eight hours), or a 15-minute short-term exposure limit (STEL), or a ceiling (C).

Primary barrier – The inner element designed to contain the cargo when the cargo containment system includes two boundaries.

Product carrier – An oil tanker engaged in the trade of carrying oil other than crude oil.

Residue – Any noxious liquid substance which remains for disposal.

Residue/water mixture – Residue in which water has been added for any purpose (e.g. tank cleaning, ballasting, bilge slops).

Secondary barrier – The liquid-resisting outer element of a cargo containment system designated to afford temporary containment of any envisaged leakage of liquid cargo through the primary barrier and to prevent the towering of temperature of the ship's structure to an unsafe level.

Segregated ballast – That ballast water introduced into a tank which is completely separated from the cargo oil and fuel oil system and which is permanently allocated to the carriage of ballast or to the carriage of ballast or cargoes other than oil or noxious substances.

Sewage – (1) drainage and other wastes from any form of toilet, urinals and WC scuppers; (2) drainage from medical premises (dispensary, sick bay, etc.) via wash basins, wash tubs and scuppers located in such premises; (3) drainage from spaces containing living animals; (4) other waste waters when mixed with drainage as listed above.

Ship – A vessel of any type whatsoever operating in the marine environment; includes hydrofoils, air cushion vehicles, submersibles, floating craft and fixed or floating platforms.

Shipboard incinerator – A shipboard facility designed for the primary purpose of incineration.

Shipboard Marine Pollution Emergency Plan (SEMEP) – Annex II of MARPOL, Regulation 17, requires that all vessels more than 150 grt carrying noxious liquid substances in bulk are required to carry a Marine Pollution Emergency Plan for Noxious Liquid Substances. These vessels must also carry an oil pollution emergency plan which contains similar guidelines. For emergency use these have been combined to make up SEMEP.

Ships Oil Pollution Emergency Plan (SOPEP) – A shipboard plan to counteract spillage of any oil discharge from the vessel. The plan would expect to incorporate cleaning materials for on board use and emergency contact numbers for designated persons ashore, to assist in counter action against any oil pollution incident.

Slop tank – A tank specifically designated for the collection of tank residuals, tank washings and other oily mixtures.

Sludge oil – Sludge from the fuel or lubricating oil separators, waste lubricating oil from main or auxiliary machinery or waste oil from bilge water separators, oil filtering equipment or drip trays.

SOx emission control area – An area where the adoption of special mandatory measures for SO_x emissions from ships is required to prevent, reduce and control air pollution from SO_x and its attendant adverse impacts on land and sea areas. SO_x emission control areas include those listed in Regulation 14 of Annex VI.

Special area – A sea area where for recognised technical reasons in relation to its oceanographical and ecological condition and to the particular character of its traffic the adoption of special mandatory methods for the prevention of sea pollution by oil is required. Special Areas include: the Mediterranean Sea, Baltic Sea, Black Sea, Red Sea, Gulf Area, Gulf of Aden, North Sea, English Channel and its approaches, the Wider Caribbean Region and Antarctica.

Substantial corrosion – The extent of corrosion such that the assessment of the corrosion pattern indicates wastage in excess of 75 per cent of the allowable margins, but remains within acceptable limits.

Suspect areas – Locations showing substantial corrosion and/or are considered by the attending surveyor to be prone to rapid wastage.

Tank – An enclosed space which is formed by the permanent structure of the ship and which is designed for the carriage of liquid in bulk.

Tank cover – The protective structure intended to protect the cargo containment system against damage where it protrudes through the weather deck or to ensure the continuity and integrity of the deck structure.

Tank dome – The upward extension of a position of a cargo tank. In the case of a below-deck cargo containment system the tank dome protrudes through the weather deck or through a tank covering.

Tanker – An oil tanker as defined by the regulation 1 (4) of Annex 1, or a chemical tanker as defined in regulation 1 (1) of Annex II of the present Convention.

Threshold limit value (TLV) – Airborne concentrations of substances devised by the American Conference of Government Industrial Hygienists (ACGIH). (Representative of conditions under which it is believed that nearly all workers may be exposed, day after day, with no adverse effects). There are three different types of TLV: time-weighted average (TWA); short-term exposure limit (STEL); and ceiling (C).

NB. TLVs are advisory exposure guidelines, not legal standards, and are based on evidence from industrial experience and research studies.

Time-weighted average (TWA) – The average time over a given work period (e.g. eight-hour working day) of a person's exposure to a chemical or an agent. The average is determined by sampling for the containment throughout the time period and represented by (TLV – TWA).

Toxic products – Those identified by a 'T' in column 'F' in the table of Chapter 19 of MARPOL.

Ullage – The measured distance between the surface of the liquid in a tank and the underside decking of the tank.

Vapour pressure – The equilibrium pressure of the saturated vapour above the liquid expressed in bars absolute at a specified temperature.

Void space – An enclosed space in the cargo area external to a cargo containment system, other than a hold space, ballast space, fuel oil tank, cargo pump or compressor room, or any space in normal use by personnel.

Volatile liquid – A liquid which is so termed is one which has a tendency to evaporate quickly and has a flash point of less than 60 °C.

Wing tank – Any tank which is adjacent to the side shell plating.

Pollutants Other than Oils

The shipping industry is constantly engaged in the movement of products considered harmful to the environment, including gases, chemicals and contaminated ballast waters.

Figure 6.2 The *SAARGAS*, a gas carrier, seen operating in the river Mersey on approach to Liverpool. The construction is distinctive from the above-deck structure of the upper part of the cargo tanks being either octagonal or a dome structure to contain pressurised liquefied gases (LNG) in bulk.

Vessels designed to carry bulk gas products are designed and constructed under stringent regulations of the IGC code. Double hull construction in way of all cargo spaces is a requirement, and primary and secondary barriers surround the cargo element.

The Causes of Maritime Pollution

Oil spills, chemical discharges and similar pollution incidents within the maritime environment are generally caused by one or a combination of the following:

- accidents during loading, discharging or transferring cargo or bunkers;
- collisions between another carrier or a fixed obstruction.
- grounding, stranding or beaching;
- fire or explosion on board;
- enemy action or sabotage.

In virtually every case the human element is usually a contributing factor.

The shipping industry, inclusive of the offshore, fishing, ports and harbour sectors, as well as salvage operations, has built-in remedies to reduce damaging effects following an incident. These take the form of corrective activity once normal and routine procedures have failed:

- internal transfer of liquids from damaged tanks to undamaged, intact tanks;
- external transfer from the carrier vessel to a lightening oil barge;
- boom and barrier equipment to prevent pollution spreading;
- skimmer vessels to scoop and recover from the surface;
- coagulation chemicals for use, assuming chemicals that are non-harmful to the environment;
- emergency equipment base stations situated at sensitive areas near shipping focal points.

The organisation of counteracting a spillage is supported by an extensive communication system to include designated persons ashore, marine pollution control units, usually established within Coastguard organisations, as well as salvage recovery operators around the globe.

Figure 6.3 A tanker approaches an oil floating storage unit (FSU). The flare boom is prominent and seen yarded well clear on its port side.

Figure 6.4 A tanker's forward mooring arrangement. The floating pipeline is seen on the surface prior to recovery by the vessel before engaging in oil transfer.

The Design of the Oil Tanker

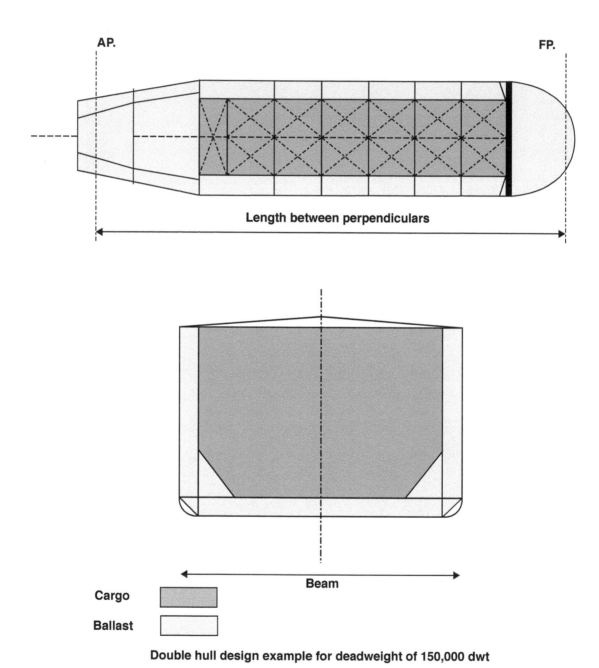

AP.

FP.

Length between perpendiculars

Beam

Cargo

Ballast

Double hull design example for deadweight of 150,000 dwt

Figure 6.5 Double hull design example for tanker of 150,000 dwt.

Oil Tankers

The very nature of the trade dictates that these vessels are large and lengthy in construction. In a light condition they are massively exposed above the water surface and can be positively influenced by strong winds. When fully loaded they have a small freeboard and are heavy to manoeuvre.

The two high-risk problems with the oil trade are pollution occurring to the environment and fire. They carry extensive pumping arrangements for conducting liquid cargo operations, and personnel on board need to be well practised in safety procedures in order to carry out acceptable operations of oil cargo movements.

Figure 6.6 A large oil tanker seen passing through the Dardanelles, en route to the Black Sea. Amidships cranes are seen at the centre of the foredeck to facilitate the lifting on/off and securing of the oil pipelines to the manifold position.

Pipeline Connections

Figure 6.7 The manifold pipe connections seen on the deck of a tanker in dry dock.

Figure 6.8 Pipeline example on the deck of a tanker.

Anti-Pollution Measures

Where manifold connections are across open decks, in routine loading/discharging procedures in any sector of the industry, shipboard procedures can very often eliminate completely or certainly reduce pollution effects.

Such anti-pollution measures would expect to follow the SOPEP, including:

- sealing the upper deck and plugging all deck scuppers;
- displaying international code signals to indicate the ship's activity, day or night;

- displaying 'no smoking' hazardous warning signs around activity areas;
- monitoring tank soundings on liquid levels;
- ensuring drip trays are strategically placed to cover pipe flanged joints;
- having an established three-way communication loop that includes the pump station, manifold and tank storage position;
- having emergency SOPEP equipment kept on immediate readiness, together with the emergency contact numbers of the designated person ashore.

On the technical side, any oil spills would be anticipated by having absorbent materials and decontamination facilities available to ensure a total clean-up. Many vessels now carry limited supplies of barrier/boom gear to enclose the immediate area of any spillage incident. However, large areas of spillage would inevitably need shoreside support in the supply of extended barrier equipment.

Selected geographic locations tend to retain emergency equipment supplies to be held readily available in the event of a pollution incident. Such equipment would be likely to contain hydrocarbon absorbent materials, barrier equipment and emergency towing vessels. Such equipment stations are available for not only the tanker sector of the industry, but also general shipping associated protection of the marine environment.

After the *Exxon Valdez* and the Deepwater Horizon incidents, authorities are more than aware of the damage that can be inflicted on the marine environment from the shipping industry. Prevention is always better than the cure.

Oil Spills

The marine environment has unfortunately experienced many oil spills over the years. Oil pollution occurs not just from tanker vessels, but from grounding and collision incidents the world over, from non-tanker vessels. Ships, for one reason or another, become involved in accidents causing harmful effects to the waters and coastlines. These are mostly the maritime countries, but also inland waters like rivers and lakes which operate marine traffic can expect to experience similar incidents.

The more recent disasters of note are: the Deepwater Horizon oil platform in the Gulf of Mexico (2010); the *Exxon Valdez* off the Alaskan coastline (1989); the Piper Alpha Platform in the North Sea (1976); the *Sea Empress* tanker off Mildford Havern (2004); the vessel *Prestige* (2002) off Spanish, French and Portuguese coastlines; The *M.V. Braer* (1993) off the Shetland Islands; The *Torrey Canyon* (1967) off the Scilly Islands, etc.

The list of disasters is long and not limited to the few that have been noted. They have all been expensive, often in the loss of life and most certainly in monetary value, with clean-ups taking months beyond the date of the casualty. The detrimental effects to the fishing community and local wildlife are of usually immeasurable proportions. Any loss of revenue experienced by the actual loss of the carrier or the delivery system is not included in what would be a hypothetical bottom line.

The fact that spills seem inevitable over an indefinite period of time has caused emergency equipment stations to be established at various locations throughout the maritime communities. These stations tend to be established in high-density traffic areas and can supply essentials like boom/barrier gear, anti-pollution chemicals, emergency tugs and skimmers, where appropriate. In virtually every incident, local manpower, communications, helicopter transports and cleaning apparatus are employed, very often against the adverse elements of the weather.

Exxon Valdez, 23 March 1989

Damage Summary following oil spill after running aground; position off 'Prince William Sound' on the Alaskan coast.

In the aftermath of the grounding incident of the oil tanker *Exxon Valdez* it was found that the damage from an oil spill of 10.8 million gallons into the sound were:

- 1,300 miles of coastline contaminated by the spill;
- as many as 2,800 sea otters died;
- 900 bald eagles died;
- 250,000 seabirds died;
- 300 harbour seals died;
- 1,000 harlequin ducks died, with many more affected;
- the clean-up operation required 10,000 personnel, 1,000 boats and about 100 aircraft, both fixed and rotary wing;
- four deaths were directly related to the clean-up efforts;
- $300 million of economic harm affected 32,000 people whose livelihoods depended on commercial fishing;
- many fish populations were harmed, including the sand lance and herring; significantly fewer quantities were landed in 1992 and 1994, with many fish having increased viral infections;
- pink salmon embryos that remained continued to be harmed by oil remnants until at least 1993. It is estimated that the region lost 1.9 million pink salmon, about 28 per cent of the total stock. By 1992 the affected part of the sound still had 6 per cent less wild pink salmon than would exist if the spill had not occurred;
- there was an estimated loss of 9,400 visitors to the area and $5.5 million paid out in state spending;

- tourist spending decreased by 8 per cent in south central Alaska and by 35 per cent in south west Alaska in the year after the spill;
- two years after the incident, the economic loss to recreational fishing was estimated at $31 million;
- 12 years after the spill oil could still be found on half of 91 randomly selected beaches. Bird and wildlife species have still not recovered numbers.

Lightening Operations (Ship-to-Ship Transfer)

Where tanker vessels are unfortunate enough to run aground, one of the solutions to refloat the vessel often employs the use of a 'lightening vessel(s)' being brought in close to the aground ship. This then allows the use of a flexible, floating oil-bearing pipeline to be connected to the stricken vessel to allow the oil cargo to be transferred. This discharge effectively reduces the draught and allows the ship to be floated off.

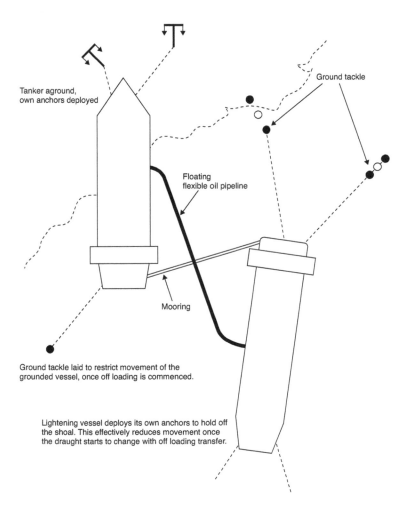

Tanker aground, own anchors deployed

Ground tackle

Floating flexible oil pipeline

Mooring

Ground tackle laid to restrict movement of the grounded vessel, once off loading is commenced.

Lightening vessel deploys its own anchors to hold off the shoal. This effectively reduces movement once the draught starts to change with off loading transfer.

Figure 6.9 Example lightening operation from a vessel aground to a relief lightening vessel. Ground tackle deployed to prevent unwanted movement to either vessel.

Ship-to-Ship Oil Transfer

A variety of cargoes, including bulk commodities and liquid cargoes, can be salved from vessels which find themselves in a stranded situation. The geographic aspects, especially the depth of water, will tend to dictate the closeness and proximity that any recovery vessel can attain in order to effect transfer. However, the use of floating pipelines can be employed to gain transfer even at extended range, for the majority of liquid cargoes.

Ideally, shuttle tankers (lightening barges) are better and more economically employed when they can get alongside. The risk of spillage is greatly reduced with ship-to-ship transfer when both carrier and reception vessel can attain alongside positions. Where solid cargoes are expected to be transferred, lifting gear is more frequently employed and the alongside position becomes essential to effect transfer within the range of the lifting gear.

In all cases of ship-to-ship transfer it would be essential to secure the position of both vessels prior to conducting actual transfer. To this end the carrier may well be fixed in an aground position, but may still be affected by tidal movement. To counter the risk of breaking away, ground tackle is frequently laid to prevent additional movement by natural, geographic elements like winds and tides. Ground tackle by way of anchors and chains can be extensive and in themselves be difficult to establish in order to reduce movement in grounded vessels. This is especially so, as when large tonnages are being removed, the ship is getting lighter and more buoyant, generating movement on the hull. Prudent ballasting as loads are removed may go some way to counteract the transfer of weights.

NB. Vessels so involved can expect to be under 'restricted in ability to manoeuvre' navigation signals, while stranded vessels would show 'aground' signals. Prudent use of the International Code of Signals should be employed, e.g. Bravo flag.

Any operation, especially one which employs floating pipelines, must be well planned beforehand and should be conducted to coincide with a period of good weather, preferably below force 5 on the Beaufort scale. Heavy, short swell sea conditions should also be avoided where possible. The risk of damage to pipelines is extremely high in bad weather conditions. Long-range weather forecasts and continual updates on local weather conditions can make the difference between a successful and an unsuccessful transfer operation.

All such operations are expected to have in place effective communications between transfer stations, mobile units and shoreside authorities. Anti-pollution equipment should be on site and readily available in the event of any pollution which may arise.

This equipment should include dispersal chemicals in sufficient quantities, together with the respective permissions for the use of the same. Log book records should be maintained throughout the operation on all relevant transports until operations are terminated. Stand-by vessels with extended capabilities should be in attendance throughout any and all operational procedures.

Recovery of Floating Oil Pipelines

Figure 6.10 Chain and stopper arrangements for secure connections to a floating storage unit.

Figure 6.11 Floating oil pipeline pick-up buoys used to recover pipelines prior to securing to a manifold connection aboard the tanker vessel.

Figure 6.12 Oil cargo pipe line connections to tanker manifold.

Figure 6.13 Oil pipe flange connection to manifold.

Oil Movement

Every movement of oil has the integral risk of pollution to the environment. It is essential that those persons involved in transfer of oil cargoes, bunker oils or similar products are conscious of the risks to the environment and take all precautions prior to movement taking place.

When engaged in the routine of taking bunkers, ships will be expected to go through and complete a checklist to ensure an incident-free operation. A typical checklist would include most of the following:

1 Upper deck areas with manifold connections must have all scuppers sealed.
2 Double access from the ship must be provided during any period of bunkering.
3 Additional no smoking signs must be displayed to relevant deck areas.
4 The ship should exhibit the 'Bravo' flag or show a red warning light while taking bunkers.
5 No unauthorised persons should be allowed on board.
6 A three-way communication link should be established between the pump station, the manifold and the tank top receiving the oil. The function of this link is to stop the flow in the event of airlock or blowback to the pipe.
7 Adequate manpower must be available, especially when topping off.
8 All connections should be fitted with drip tray coverage.
9 All SOPEP equipment must be made readily available prior to commencing bunker operations.
10 The emergency contact numbers for the designated person ashore should be made readily available.
11 All fire precautions and conventional apparatus should be on site at the manifold.
12 Soundings of the receiving tank should be taken before, during and after loading bunker oil.
13 Statements of starting and completing bunkers should be entered in the oil record book and in the ship's deck log book.

The Master must be informed of this activity before bunkering commences. It would be normal practice for engineers to test the grade of oil and check quantities and temperatures with shoreside personnel. On completion of bunker operations, the total quantity in tonnes must be passed to the ship's Chief Officer to allow a revised stability assessment to be made.

Incident Report: Grounding of the Drilling Rig *Kulluk*, 30 December 2012

The Royal Dutch Shell drilling rig *Kulluk* broke away from its towing lines in bad weather off Kodiak Island in Alaskan waters. It was reported that the drilling rig was carrying 143,000 gallons of diesel oil and 12,000 of other oil products. The *Kulluk* was constructed with a double sided hull of reinforced steel, but the vessel still experienced heeling from side to side in heavy seas before grounding.

A six-man assessment team was landed by a USCG helicopter on 2 January 2013, with a state-owned emergency towing system. Smit Salvage headed up the operation to free the rig and attempted to refloat the unit. The rig was driven ashore in storm force weather conditions.

To date, no pollution of the environment has occurred.

Such an incident, inside the Arctic region, illustrates that the combination of bad weather in a particularly sensitive area could and does happen. Many marine disaster incidents are associated with bad weather, which frequently escalate minor incidents into major catastrophes, not always with conventional shipping but with associated affiliates of the offshore or fishing fraternity.

In the case of the *Kulluk* incident, the towing vehicle lost power in bad weather, leading to the parting of the towline. Escalation was ongoing, with continued bad weather in the area into the new year.

This particular incident is somewhat notable with the Arctic region opening up to extensive offshore development to tap into the natural resources of the hostile environment. Obvious concerns preside over the pristine condition of the Arctic and Antarctic regions to protect the environments from the effects of pollution. Public concern is being continually expressed by environmentalists of Greenpeace and the Wilderness Society.

Oil Recovery Equipment

The vessels are designed and adapted to handle, store and transport oil recovered from a spill in emergency situations. These vessels are expected to comply with the 'Code of Practice for Vessels Engaged in Oil Recovery Operations'. They are classed as tankers within the meaning of the Merchant Shipping Regulations, but it is recognised that these vessels may not be able to meet all the tanker requirements.

These vessels are inspected by the Marine Authority and, provided they are found satisfactory, are issued with an Oil Recovery Certificate. This provides a general exemption from regulations pertaining to tankers and allows the vessel to operate

Figure 6.14 Oil recovery equipment seen on board a designated oil recovery vessel. It is used to recover surface oil and separate it from water content and other foreign bodies, to permit the product to be recycled. Use of skimmer vessels can be extremely effective in the reclamation of oil spillage over large areas of water.

as a dedicated oil recovery vessel (the period of validity of the certificate is two years).

Ballast Water Movement

Ballast water management is the means – mechanical, physical, chemical or biological – either singularly or in combination to remove, render harmless or avoid the uptake or discharge of harmful aquatic organisms or pathogens within ballast water and sediment.

It is now a requirement, since ratification of the Ballast Water Convention, that all ships operate with a Ballast Water Management Plan which must incorporate records of water exchange operations. Ship inspectors will expect to see a ballast water exchange report which stipulates the location, date and time of all ballast water exchanges made to the ship.

Control of ballast water has come about in order to prevent the import of non-native aquatic organisms into geographic areas that would experience detriment from their addition. Notable areas requiring protection have been places such as the Great Barrier Reef of Australia, the Mediterranean and Caribbean Seas, where excessive growth of weed has occurred due to foreign organic matter being deposited outside of its natural habitat.

The IMO regulations now require commercial vessels to have a ballast water treatment plant on board. The function is to effectively filter all water ballast by any acceptable and suitable means. This might include chlorination, disinfecting or mechanical filter apparatus. Some units pass water through ultraviolet light to ensure organic matter is eliminated. The outcome is to deliver clear ballast waters back to the oceans and waterways of the marine environment.

7

Towing and Salvage Hazards

Introduction

Many of the incidents covered by the previous chapters – collision, grounding, lee shore scenarios, fire, etc. – are often followed by the involvement of tugs, either as a towing operation or in a stand-by, supporting role. Clearly a towing contract is the more desirable arrangement as opposed to a salvage operation, but choice may be limited by the circumstances at the time.

Emergency towing arrangements for tankers have been around since 1987, but more recent legislation that came into force in January 2012 has made emergency towing arrangements (ETAs) statutory for all commercial ships over 500 grt.

Engaging with tugs for docking purposes is not an unusual task, but long-range towing operations are not generally considered as routine for general shipping. Oceangoing tugs are employed for long-distance towing operations. The endurance of a heavy-duty tug is far greater than the harbour control tractor tugs used for routine docking procedures. Oceangoing tugs, prior to contract, would probably undergo survey/inspection as to their capabilities and capacity to complete the task. A 'warranty surveyor' is normally employed to assess the vessel and the overall towing operation.

Supervision and advice on securing towlines will tend to be offered by the 'towmaster', who may or may not be the Master of the towing vessel. Towlines after the unexpected incident tend to employ the ETA or a composite towline arrangement. Where a long-distance towing operation is pre-planned, as with taking a vessel for scrapping, a towing bridle arrangement may by employed as an alternative to an ETA.

Major incidents in commercial shipping frequently need other support craft in the way of floating cranes, high lift-capacity sheer legs or flat-top transport barges. Such items of equipment are available from international salvage companies. Their specialised equipment played active roles in the recovery of the *Herald of Free Enterprise* and the *Tricolour* disaster, both in the English Channel.

Tug Operations

There are many variations in the use of tugs. Tractor tugs tend to dominate the ports and harbours of the world and are well known for accompanying the larger vessels into berths and moorings. They can be employed as a 'braking' tug, providing a reduction to forward momentum. Alternatively, they can be designated in steerage and control from the bow or stern positions of the parent vessel. For disabled vessels they act as the driving force to push or pull and provide the overall speed over the ground.

Outside of harbour limits the oceangoing tug is available for salvage operations or assisting the vessel in distress.

Figure 7.1 Passing and setting up the towline from the amidships towing position on the after deck of a medium-sized tug. The deck shows the additional option to use the towing hook by way of the towing winch and use of a towing spring hawser.

Plate 1 The *Costa Concordia* cruise ship seen lying on its starboard side after striking the rocky outcrop of 'Isola del Giglio' off the Italian West Coast. The vessel is seen shortly after grounding and partial capsize with an anti-pollution boom/floating barrier surround.

Plate 2 The large container ship *MSC Luciana* (11,600 TEU) aground in September 2011, on the banks of Weilingen-Scheldt River. The vessel taking the ground on a high tide is seen assisted by nine tugs in the attempt to refloat. The vessel is 362 metres in length and has a gross tonnage of 131,771, one of the larger container ships of the day.

Plate 3 The bow damage to the Liberian registered *ER Hamburg*, following collision in the waters of the Greek Islands. The vessel is under tow by means of a chain bridle and tug escort. The towing operation is being conducted by the Tug 'Megas Alexandros'.

Plate 4 The *MV Elli*, seen afloat from a partially submerged stern. The crude oil tanker broke her back while anchored off Suez, in August 2009, while in ballast. The 94,000 dwt ship was then beached at Suez, cut into two pieces and each section was towed to Aden.

Plate 5 The Liberian registered *Akti N*, a Chemical and Products tanker ran aground in June 2009, at the Flushing Boulevard. The ship is seen under floodlights with five tugs in attendance, and civilian observers watching the attempt to re-float the vessel.

Plate 6 The container vessel *Expedition Flaminia 2* of the Mediterranean Shipping Company, seen on fire in the forepart, being assisted by the support vessel *Fairmount Expedition*. The exact cause of the fire is unknown.

Plate 7 The container ship, *MV Hansa Brandenburg* seen after a container fire caused abandonment of the vessel in the Indian Ocean, North of Mauritius, July 2013. The crew were successfully evacuated after the fire spread to the superstructure. The vessel was then towed to Port Louis, Mauritius, where the damage visible to the bridge front identifies the site and extent of the fire.

Plate 8 The dry cargo vessel *Inspiration 1* seen well aground on a rocky foreshore. The starboard anchor has been deployed and the aground signals are displayed from the bridge. Access is by a pilot ladder and one of the forward cranes has been topped for loading essential stores and equipment.

Plate 9 The cargo vessel *Western*, seen under tow in rough sea conditions. Difficult conditions with excess slack water on the after deck of the support/supply vessel which shows four crew members tending to the towline. The men waist deep in water at times, are seen working with security life lines.

Plate 10.1 The bulbous bow section of the container vessel *Cafer Dede* aground against a rock bluff shoreline South East of Syros Island in the Greek islands. The grounding incident occurred in November 2011, and the ship was re-floated 11 days later, after a partial discharge of some of her 800 containers.

Plate 10.2 Inflatable life rafts seen aboard a passenger cruise ship. The life rafts are secured by web straps and fitted with disposable 'Hydrostatic Release Units'.

Such life rafts are additional to lifeboats and davit launched life rafts in the majority of cases to satisfy the requirements of the Life Saving Regulations of the Marine Administration.

Plate 11 The *Riverdance* seen where the tidal water has receded. Some drop trailers are still seen under the accommodation structure, on the upper vehicle deck. Image courtesy of Mr George Edwards. Marine Engineer.

Plate 12 The Class 1 cruise ship, *Queen Victoria* seen manoeuvring with tug assistance through Southampton water. The view reflects the size and height of the multi-decks over and above the boat deck.

Ships of this size are designed to accommodate increased numbers of persons who may be only briefly familiar with limited deck areas. Although fire fighting and lifesaving appliances are in abundance inside passenger ships, an incident may generate complications with increased numbers of deck levels and the survival craft positioned at what is considered an acceptable freeboard, to permit safe launching.

Harbour and Port Authority Tugs

Figure 7.2 Example of a 'tractor tug', seen lying alongside in port. This size of towing vessel is generally employed in coastal waters to assist in the berthing and control of the larger oceangoing vessels when inside rivers, estuaries and harbours.

Oceangoing Salvage Tugs

The larger, longer range tug is often multi-functional by way of acting as a support vessel, anchor handler, ice breaker, etc. It is a large towing vessel with a high bollard pull (BP). They generally have designated towing winches and the capability to recover the disabled or distressed vessel following loss of machinery power or steerage.

The usual function of an oceangoing tug tends to be towing operations to deliver vessels to repair facilities, such as dry docks and dockyards. Alternative employment is taken up with towing dumb lighters and cargo barges which have no self-propulsion capabilities. Considerable activity for the large towing contractor is also found in the offshore oil and gas industries and associated civil engineering projects. Towing mobile 'offshore installations' and project cargoes, especially within developing fields, is not uncommon. For example, a recent (2008) long-distance towing example was seen with the sale of a floating dock from B & V in Germany to the Ship Repair yard in the Bahamas.

Where an oceangoing towing operation is to take place, the tug would normally be contracted by the owners or managers of the project. They would look at the capabilities of the vessel with regard to fuel and water endurance, in addition to the obvious Bollard Pull (BP). To this end the size of the tug and the length of the voyage will be influencing factors.

A towmaster would be appointed, as would a warranty surveyor to oversee the requirements of the insurer accepting the risk. Although the towmaster and the warranty surveyor may work closely together, their interests may be different in the venture.

The Work of the Towmaster

Prior to any outward tow operation, a series of meetings will take place between respective parties. The owner/contractor may be present, along with the appointed towmaster, the respective tug-master(s) where more than a single tug is engaged. The warranty surveyor would also attend procedural meetings.

The respective tugs would have undergone detailed inspections to ascertain that their certification is in place and functional in relevant detail. Every vessel engaged is assessed as to its capabilities, by both the towmaster and the warranty surveyor. In particular, the BP capability to operate at the designated speed would need to be seen to be in place.

An agreement of the endurance capabilities by way of stores, fuel/water capacity and the amount of bunkers already on board would be agreed. The passage plan would be examined in depth and refuelling destinations (if necessary) would be identified. The passage plan would need to be enhanced by appropriate contingencies where extenuating circumstances may prevail while en route, and these would be identifiable by operating personnel.

The towmaster would normally inspect the main towing gear and the emergency towing gear prior to the initial meeting. This would entail the inspection of all winches, bridles, pick-up buoy and suitable tail length with its respective break-load.

Additional equipment items required would also be sighted and seen to be in good order:

- portable generator of 35–60 kW, capable of supplying power to submersible pumps;
- portable compressor with an estimated output of 250 cu.ft/min;

- emergency lifesaving appliances;
- emergency communication equipment such as signalling equipment and walkie talkies;
- suitable emergency boarding gear, e.g. rope ladders, scrambling nets, etc.;
- emergency repair materials such as steel plate, sand and cement, plywood, hook bolts, etc.;
- appropriate portable pumping equipment;
- appropriate navigation lights and special signals.

> **Comment:** It should be borne in mind that the remoteness of a casualty's location may not lend itself to the high level of detail of a purpose-planned contract of tow agreement. As such, jury rigging of towlines, making use of anchoring systems, may be appropriate.

Tugs and Emergency Towing

Whenever deep-sea ocean towing is undertaken it is invariably non-routine and very often follows breakdown or incident that could generate salvage. The mere acceptance of the tug's line has become tacit agreement between parties that the services of the tug have been accepted, and as such this could lead to a possible unnecessary claim for salvage.

The use of tugs in salvage is well known, but the type of vessel being towed will vary and generate a specific type of towline. Most large commercial vessels would probably opt for a 'composite towline', which is comprised of the ship's anchor cable joined to the towing spring of the tug. This is a comparatively easier option for the crew to establish than, say, a bridle arrangement. However, a large tanker would be expected to engage the ETA, a required fitment with which all tanker vessels over 20,000 grt are expected to be equipped.

Recent legislation instigated from 1 January 2010 now requires all vessels over 500 grt and passenger vessels carrying more than 12 passengers to be fitted with an ETA in the fore and aft positions of the vessel (similar to the tanker requirements of 1983). Such arrangements, once fitted, will clearly supersede the use of a composite towing arrangement, which has previously been widely employed with shipping casualties in the past.

Other towing arrangements where time, manpower and equipment are available may give rise to a bridle arrangement, but this would be generally difficult to set up in open-sea conditions. If the situation warranted an alternative arrangement for towing, the possibility of engaging two or more tugs, secured from different positions, might prove a superior option.

In the majority of cases where a dedicated towing operation is to take place, it would be normal practice for a 'tug approval survey' to be conducted.

Tug Approval Surveys

A tug approval survey would be carried out by a warranty surveyor and he or she would without doubt employ a checklist to ensure the tug meets the requirements of the task in hand. Any defects found aboard the towing vessel would be passed to the base office of the surveyor. Here, a full evaluation of the defect would be made and an appropriate line of action suggested to the principal client.

The capacity of the tug and capability of the engines to deliver the required BP come under close inspection by the surveyor, as would its towing equipment, condition of hooks, winches and relevant deck equipment

In the event that the scope of the operation involves 'barge transport', then a barge inspection/survey would also be conducted. Such an inspection would pay particular attention to the watertight and structural integrity of the barge unit. Special attention would be given to the towing arrangement of the barge and any fitted equipment to be employed in the operation. The certification of the barge would also be checked and seen to be in good order.

Where cargo is to be loaded to a barge transport, this would undergo a 'cargo securing inspection'. Operational surveys would take account of the loaded barge and its draught. This feature would be critical to the route planning to ensure adequate under-keel clearance throughout the period of transport, taking into account the time period of the voyage and the tidal data relevant to geographic waypoints.

NB. The minimum tug power envisaged would be calculated against a 'design storm', the criteria of which would be expressed in terms of wind speed, wave height and speed of surface currents.

Statutory Emergency Towing Arrangements for Large Tanker Vessels

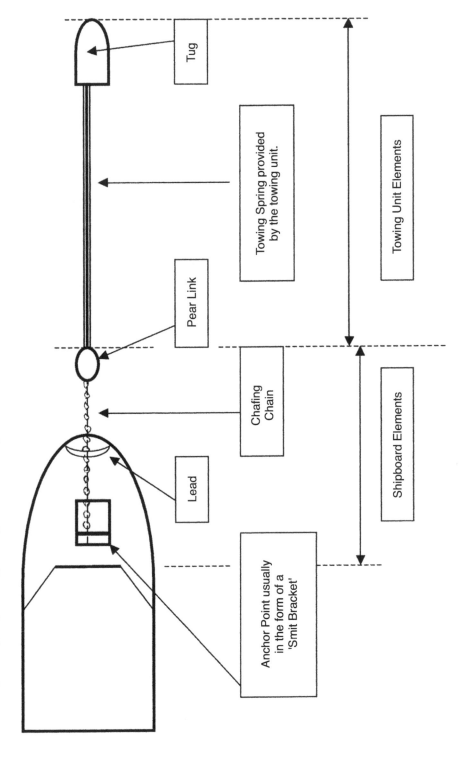

The shipboard elements of the emergency towing arrangement include the anchor point securing, a lead to accept the chafing chain and the pear link. The chafe chain is generally stowed in a box arrangement to one side of the forecastle head from where it can be easily deployed. Alternatively some vessels may carry their own towing spring unit below decks which can be easily shackled to the chafe chain and deployed through the purpose designed lead.

Figure 7.3 [To be completed at proof stage]

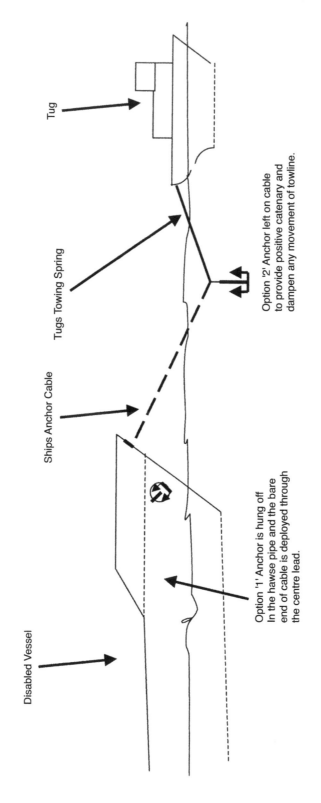

Composite Tow Line

Disabled Vessel

Ships Anchor Cable

Tugs Towing Spring

Tug

Option '1' Anchor is hung off
In the hawse pipe and the bare
end of cable is deployed through
the centre lead.

Option '2' Anchor left on cable
to provide positive catenary and
dampen any movement of towline.

Length of towline can be adjusted easily
by the disabled vessel, if manned.

(2nd Anchor of disabled vessel is left ready for emergency use)

Figure 7.4 [To be completed at proof stage]

Cargo Deck Barges (Pontoons)

Barge transports are frequently used within the salvage industry and many of the main salvage operators tend to have their own craft in this category. The more sophisticated barge transports are equipped with motive power and means of steerage; however, tug support vessels are often used in barge movements.

Barge transports can be multi-purpose, and to this end they are usually provided with a variety of specialised equipment, which could include a selection or all of the following: anchors and anchor warps; anchor winch; deck recovery winch; main and secondary towing bridles; bilge pumps and tank pumping systems; ventilation and air pipes; mooring pipes and bollards around the perimeter; towing fitments; navigation light fittings; and portable gangway.

Their use can be versatile in terms of landing heavy loads for transport purposes. They have also been used as mooring platforms (pontoons) to mount additional winches and haulage equipment when open water is predominant around a particular casualty operation. Additional uses are in transport of very large project cargoes or to provide added buoyancy support to the loaded pontoon platform carrying the load.

Overall measurements and deck load capacity vary across barges. However, they all perform the basic function of being able to carry heavy equipment to an operational site and also carry away any salved commodities, inclusive of steel ship sections after cutting operations have taken place. The size of the task dictates the operational needs and respective size of the barges employed.

The Insurable Risk

Outside of harbour limits the oceangoing tug is available for delivery of project cargoes, barge transport and salvage operations. Additionally, they are often employed to assist vessels in distress, either directly or indirectly in a stand-by capacity.

In many of their functions an insurable risk is involved within the marine insurance market. The 'underwriters' are the parties who take the risk on and set the premium for a particular policy. The 'broker' will advise the client on the type of policy required and also of the requirements of the underwriters in order to set the policy for the insured risk.

A warranty survey company will often be approached by the owner/client for the task. The underwriters may recommend or impose a warranty survey company to be engaged as a pre-condition prior to instigating the policy. To this end an exchange of information between the client's brokers and the warranty survey company regarding policy conditions and specific guidelines can be

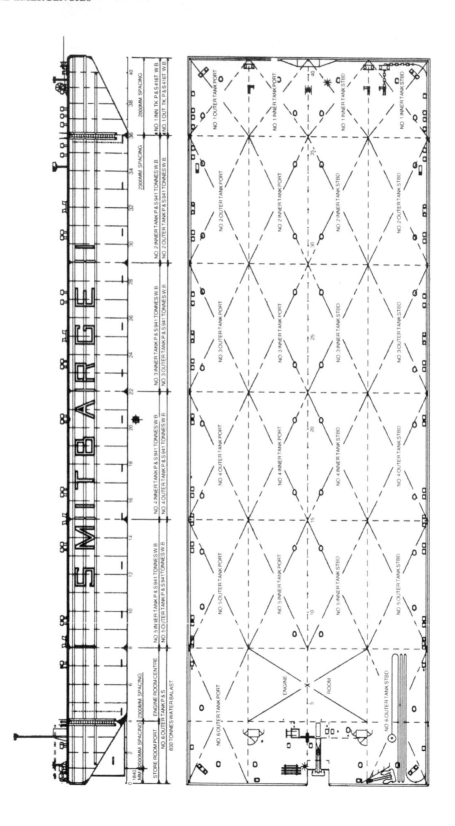

Figure 7.5 Example cargo/general purpose barge.

GENERAL PARTICULARS

Figure 7.6 General particulars for example pontoon/barge.

BUILT BY: AUSTIN AND PICKERSGILL LTD., SOUTHWICK SHIPYARD, SUNDERLAND.

LENGTH EXTREME 91.715M	LENGTH MOULDED 91.440M
BREADTH EXTREME 30.755M	BREADTH MOULDED 30.480M
DEPTH EXTREME 7.647M	DEPTH MOULDED 7.620M
SUMMER DRAFT EXTREME 6.163M	DEADWEIGHT AT SUMMER DRAFT 13980TONNES
LIGHT DRAFT 0.961M	LIGHTWEIGHT 2262TONNES

REGISTERED DIMENSIONS

LENGTH . . . 87.800METRES   BREADTH . . . 30.480METRES   DEPTH 7.620METRES

TONNAGES	DUTCH	SUEZ
GROSS	5688	18859.91 CU.M.
NET	1706	18859.91 CU.M.

CLASSIFICATION SOCIETY

LLOYDS REGISTER OF SHIPPING CLASS + 100A1 'PONTOON'

TANK CAPACITIES

Figure 7.7 Tank capacity and disposition table for example barge.

COMPARTMENT		FRAMES	VOLUME CU. MTRS	W.B.	F.W.	F.O.	L.C.G. FWD. AP METRES	V.C.G. ABOVE BASE METRES
					TONNES			
NO.1 INNER TANK	PORT	36–41	406	416	406		85.697	4.727
NO.1 INNER TANK	STBD	36–41	406	416	406		85.697	4.727
NO.1 OUTER TANK	PORT	36–41	406	416	406		85.697	4.727
NO.1 OUTER TANK	STBD	36–41	406	416	406		85.697	4.727
NO.2 INNER TANK	PORT	29–36	918	941	918		73.390	3.810
NO.2 INNER TANK	STBD	29–36	918	941	918		73.390	3.810
NO.2 OUTER TANK	PORT	29–36	918	941	918		73.390	3.810
NO.2 OUTER TANK	STBD	29–36	918	941	918		73.390	3.810
NO.3 INNER TANK	PORT	22–29	918	941	918		57.290	3.810
NO.3 INNER TANK	STBD	22–29	918	941	918		57.290	3.810
NO.3 OUTER TANK	PORT	22–29	918	941	918		57.290	3.810
NO.3 OUTER TANK	STBD	22–29	918	941	918		57.290	3.810
NO.4 INNER TANK	PORT	15–22	918	941	918		41.190	3.810
NO.4 INNER TANK	STBD	15–22	918	941	918		41.190	3.810
NO.4 OUTER TANK	PORT	15–22	918	941	918		41.190	3.810
NO.4 OUTER TANK	STBD	15–22	918	941	918		41.190	3.810
NO.5 INNER TANK	PORT	8–15	918	941	918		25.090	3.810
NO.5 INNER TANK	STBD	8–15	918	941	918		25.090	3.810
NO.5 OUTER TANK	PORT	8–15	918	941	918		25.090	3.810
NO.5 OUTER TANK	STBD	8–15	918	941	918		25.090	3.810
NO.6 OUTER TANK	PORT	0–8	810	830	810		9.658	4.227
NO.6 OUTER TANK	STBD	0–8	810	830	810		9.658	4.227
FUEL OIL TANK	STBD	3–4	26			23	6.840	4.851
TOTALS			17932	18380	17932	23		

W.B. . . . . . . WATER BALLAST . . . . 0.9756CU.MTRS./TONNE . . . S.G. 1.0250
F.W. . . . . . . FRESH WATER . . . . . 1.0000CU.MTRS./TONNE . . . S.G. 1.0000
F.O. . . . . . . FUEL OIL . . . . 1.1150CU.MTRS./TONNE . . . S.G. 0.8969

DECK LOADINGS

LOCATION	LOADING
MAXIMUM SURFACE LOADING	15 TONNES/SQ.MTR.
CROSSING POINTS OF BULKHEADS	435 TONNES
CROSSING POINTS OF BULKHEADS AND TRANSVERSES	260 TONNES
BULKHEADS AND TRANSVERSES	160 TONNES

Figure 7.8 The open aspect of the aft deck area of the oceangoing support vessel. A multi-purpose vessel, it is employed for anchor handling, buoy or navigation mark removal, as well as long-range towing.

expected. A warranty surveyor will undoubtedly be appointed to act in the field and be aware of the client's proposed procedures to ensure they comply with recommended guidelines.

Sheer Legs in Salvage Use

Many methods of salvage have involved lifting vessels or part vessels from the regions of the sea bed. Floating barge cranes and, more recently, multiple 'sheer legs' have been more gainfully employed. These mammoth structures have long seen established work in and around the offshore sector in the oil/gas industries and in civil engineering projects, bridges and harbour facility construction.

Where incident vessels have been cut into sections, sheer legs, with lifting capacity in excess of 3,000 tonnes, are frequently engaged in recovering the large and heavy ship parts to the surface. They often work in conjunction with deck salvage barges and tug support.

Figure 7.9 The *Smit Cyclone* floating sheer leg heavy lift barge, seen fitted with the extended jib in place, making a typical heavy-lift operation.

Many crane barges and sheer leg platforms are self-propelled, but others are moved and manoeuvred to sites by accompanying tug support craft.

Salvage Contact

Every salvage casualty generates a first contact and from this moment onwards information regarding the incident needs to be in detail and accurate. Many first contacts may be generated by the ship's Master or Officer in Charge, and the initial information passed over could become critical in the difference between saving or losing the vessel. The first contact, although often from the casualty itself, may be relayed by a third party and it becomes vital that any relayed information is exactly word for word as initially reported.

An example ship's Master's reports should contain the following:

- *Current situation report*: A statement of fact defining the character of the emergency, e.g. grounding, fire, collision, mechanical failure, together with any active countermeasures that have been initiated to relieve the situation. To include whether the ship is expected to remain afloat, sink or can be beached. The weather affecting the immediate vicinity, would need to be updated.
- *Vessel's details*: To include the ship's name and official number, together with the length O/A, maximum breadth, current draught, type of cargo and quantity. If the vessel is damaged, listed (port or starboard), or on an even keel condition and whether the vessel's watertight integrity has been compromised.
- *Condition of propulsion, power and steering*: A statement as to the status of the ship's main engines, its power supply and the condition of steering.
- *Position and movement details*: The position of the vessel in latitude and longitude, with range and bearing from a prominent landmark, if appropriate. The course and speed of own vessel (if applicable). To include also the port from which the ship has departed, the port to which she is bound, where the vessel is stopped, the drift rate in knots and the direction of the drift. The depth of water should also be included, together with the options for the use of anchors.
- *Personnel involved*: Details of total numbers of personnel on board own vessel, together with lists of injuries or fatalities. Where passengers are involved a passenger list should be included. Where appropriate, next-of-kin details may need to be communicated.
- *Pollution aspects*: A statement of fact as to whether the bunker tanks have been broached and the estimated extent of pollution present, including the type of oil lost overside, the number of tanks involved, the location of these damaged tanks and the quantity of content remaining. In addition, whether the vessel retains transferring capability and to which available tanks can she effect transfer. Whether hazardous goods are being carried on board and whether pollution has yet occurred from these cargoes. Where hazardous goods are involved these should be accompanied by the full details, including chemical names and UN numbers, together with the quantity as per the cargo manifest and/or stowage plan.
- *Communication*: Full communication detail should be provided, to include call sign, sat-coms and cell phones. With the advent of smartphones it is advised that communications from live feed may be accessed by shoreside media organisations. Any calls should therefore be guarded against open disclosure.

The initial communications is not exhaustive and additional information respective of ship type, nature of injuries and cargo specifics may also be beneficial in effecting relief. Also of note are:

- *Communication contacts*: These will undoubtedly be established between the ship's Master and the salvor, but could be expected to also include contact with the ship's owners, insurance brokers, hull and machinery underwriters, the Protection and Indemnity Association and liability insurers. Contact with salvage engineers of the Classification Society and shoreside national authorities, local authorities, legal representatives and agents of these could well have a direct input and influence use of visual signals to other traffic in the vicinity.
- *Media contact*: Virtually all major incidents and some minor operations attract media coverage. A suitable defined response should be jointly agreed between affected parties. A prepared statement is usually considered appropriate. The extensive use of smartphones within the media and the ability to access live feeds should be realised and confidential data should not be openly discussed unless a secure line of communication is available.
- *Weather reports*: It must be anticipated that regular checks on the weather affecting the area of incident would be of particular interest, together with any tidal times and effects.

Figure 7.10 All-purpose oceangoing support vessel the *Pacific Retriever* demonstrating its anti-fire capability, displaying her hose monitors.

Quality of Information

As and when communication between the casualty and the salvor takes place, the detail and quality of information reported influences the type of salvage vehicle(s) and the equipment required for a positive response. High-quality information can save time, lessen mistakes and very often save lives.

Sight of the ship's plans – i.e. general arrangement, tank disposition, rigging plans and docking plan – can be useful regarding the interpretation of relevant information supplied by affected parties. Where the ship's tanks are involved within the casualty, it would be useful to know the capacity and their contents, together with pumping arrangements available, if any. Also helpful are: what the last sounding was; the location of the tank access points; and quality of content.

Where damage to the vessel is a consideration, as in the majority of cases, the type, size and the position, in relation to the waterline, could prove helpful to a salvor when determining appropriate action. Consideration to effect either on-site repairs or establish recovery operations would also usually be influenced by the geographic location. Particularly, depth of water on site and the proximity of the nearest land, together with port or harbour facilities, in the location area.

Operation timings will always be influenced by the weather conditions, and once a casualty has been declared, normal practice would envisage going through a detailed planning stage. The meteorological aspects can clearly be expected to become a major consideration prior to operational activities taking place.

Communications will become a high priority. Names, addresses and contact numbers, together with radio operating channels, need to become widely available to affected parties. A salvage Master or marine surveyor could expect to be in communication with government organisations as well as local communities, technical operators, maritime bodies and legal and company representatives. There may also be communication with on-site recovery, working units like tugs, oil barge personnel, ships agents, etc.

At the end of the day, any operation will come down to money. Detailed records must be kept of all operations. Not only all expenditure, but of all telephone calls, radio messages, e-mails and hard-copy communications. Working hours, labour counts, equipment use and external inputs must all be paid for. Such records will greatly influence financial settlements of the final outcome.

8
Miscellaneous and Routine Leading to Potential Hazards

Introduction

No text can encompass every hazard likely to be encountered on the decks of ships at sea. One of the most detrimental attitudes towards working at sea is probably complacency in routine sea board tasks. Typical situations like the regular act of maintaining a lookout when signs of ice or fog are encountered and unexpected target proximity can cause high anxiety.

Similarly, the regular painting of ships rails, in itself, is not an emergency, but a short lapse of concentration could easily turn the scenario into a man overboard situation.

In such a situation a rescue boat crew is required to launch the fast rescue craft (FRC) and a full bridge team is established, with the ship going to an alert status. Another situation is the unexpected need to enter a tank. It is only a quick in-and-out, but the 'permit to work' is not completed fully, the necessary precautions are not taken and suddenly crew members are into a full breathing apparatus (B/A) recovery operation.

Poor judgements and simple mistakes on routine tasks escalate into hazardous situations. Is it through lack of training, lack of thought, complacency or just not applying common sense. Whatever the reason, minor infringements can expand quickly inside the marine environment. It is for this reason that continuous training is expected within every task at sea, to attain a level of preparedness for the unexpected.

Enclosed Space Entry

Mariners are regularly tasked to enter enclosed spaces for one reason or another. It is an unfortunate situation that the industry loses people regularly through individuals not following the expected safety procedures of how to enter such spaces correctly. What should be a routine practice, covered by a checklist, frequently turns into an emergency recovery exercise.

Persons entering tanks, void spaces, pump rooms, chain lockers, etc., are expected to follow a 'Permit to Work' system to ensure safe entry and exit within a potentially threatening environment. The permit is effectively a checklist to ensure the dangers of enclosed spaces are avoided.

If B/A is needed and required for entry, all checks and precautions must be taken prior to actual entry into the space. In addition to providing adequate illumination within the space to be entered, the area should also be well ventilated prior to persons moving into the space. A stand-by man should be engaged at the entrance of the space to guarantee a safe exit of all persons leaving the space.

The function of the stand-by person should not be confused as to being a rescue body in the event that the person inside the space encounters problems. The function of the stand-by man is purely to raise the alarm, which then permits a designated rescue party to effect recovery of the person(s) inside the space. Should the stand-by man enter the space as a rescue individual, he/she may equally succumb to the same problems inside the space, so doubling or increasing the casualty numbers.

The routine practice of enclosed space entry has become the emergency incident that could have been easily avoided if the person entering the space had adhered to basic training. All new entrants to the industry are informed of enclosed space entry and the dangers associated with the same. They are also introduced to the use of B/A at an early stage in basic firefighting training. If lazy practice or complacency is allowed to creep into any ship, there will always be a real threat to our mariners.

Checks to the B/A include:

1 Ensure that the gauge is reading in the 'green', meaning the air bottle is full.
2 Listen for the low-air alarm whistle when turning the air bottle valve on.
3 Tension up the mask to ensure a gas seal around the airways of the wearer is achieved.
4 Once the mask seal is established, briefly shut off air supply valve and assess the wearer for loss of breathable air. If the mask is seen to crush or collapse on the wearer's face, then he/she is obtaining no air from under the mask seal, so a gas seal is established.

Fog Encounter

Poor visibility is a condition that occurs frequently in certain areas of the world. The Grand Banks and the English Channel being well-known focal points for traffic and it is certainly not unknown to experience conditions of reduced visibility. The action taken by the Officer of the Watch could well be the action that keeps the ship safe, but equally so the lack of action by the Officer of the Watch could be what generates that collision emergency.

Most companies' and Masters' standing orders take into account that the vessel might enter an area of restricted visibility, either at night or during daylight hours. Definitive procedures are described within the Regulations for the Prevention of Collision at Sea, and also stated in the Watch Procedures Manual. There are standard actions that must be followed:

1 The ship's engines should be placed on stand-by and speed reduced (where the vessel is operated by bridge control stand-by engines may not be applicable).
2 Master to be informed of the change in visibility conditions.
3 Sounding of appropriate fog signals should be commenced.
4 Extra lookouts should be posted.
5 Manual steering should be engaged.
6 Navigation lights should be exhibited.
7 Radars should be reduced to anti-collision range and retuned.
8 All radar targets should be plotted and signatures analysed.
9 Bridge wing doors should be opened to pick up sound signals.
10 All contradictory noise on deck should be silenced.

Also, when encountering fog conditions, relevant log book entries should be made for entering and leaving the area of restricted visibility.

Additionally, depending on circumstances, the ship's Master may find it necessary and prudent to double watch-keeping personnel, and/or provide a continuous radar watch. Unmanned engine rooms may expect to be manned by duty engineers.

Many of the collisions at sea occur in poor visibility, with prevailing fog. Use of radar is recognised as being an essential element of effective watch keeping, in addition to ECDIS if carried, but has the danger of over-reliance by the Officer of the Watch. The effectiveness of the lookout, by all available means, includes radar, but also by visual and audible senses.

Some of the larger ships are fitted with forward-looking cameras, while ships over 300 grt on international voyages and cargo ships of 500 grt not on such voyages must be fitted with automated identification systems (AISs). It should be noted that an AIS is not a regulatory fitting for vessels under 300 grt.

Dangers Associated with Restricted Visibility

Dusk is a period of the day when the light is fading. Coupled with fog, visual bearings become impossible and shoals are in close proximity to the land. The need for primary and secondary position fixing systems becomes paramount when making a landfall or approaching shallow regions. The corroboration of echo sounding is an essential use in effective position monitoring of the vessel.

Poor visibility is defined by the following scale:

Scale Number	Description	Target invisible at
0	Dense fog	45 metres
1	Thick fog	1 cable (183 m)
2	Fog	2 cables
3	Moderate fog	0.5 miles
4	Mist or haze	1 mile
5	Poor visibility	2 miles
6	Moderate visibility	5 miles
7	Good visibility	10 miles
8	Very good visibility	30 miles
9	Excellent visibility	

Doubling Watches

Certain conditions, like when a ship encounters fog or dangerous ice formations, may cause the ship's Master to instigate a double watch routine. Establishing double watches effectively means increasing the work load and duty time on Watch Officers. The procedure has with it the inherent danger of generating fatigue among duty personnel and should not be adopted lightly.

Many cases of fog or ice encounters do not immediately justify installing a double watch routine. Prevailing circumstances at the time will influence any decision by the Master. Clearly, the geography and the expected quantity of traffic should be considered.

An example is fog in the English Channel. Doubling watches at the Western Approaches might be desirable, but would probably leave Watch Officers tired at the Dover Straits. Doubling watches abeam of the Isle of Wight would have less chance of having tired men when nearing the critical area of the Dover Straits. Tired watch-keepers, make mistakes and when the ship is at a focal point of traffic is when the Master wants all his officers sharp and alert.

Conditions might be such that a Master needs a constant radar watch to be maintained. This is typically inside ice limits, high latitudes, winter months and with thick fog prevailing. The prudent Master might feel this is the time to double watch-keepers. It will not be the most popular decision on the ship, but the Master is not running a popularity contest. His only concern is to keep the ship and all on board safe. If that means double watches, so be it.

From experience, double watches tend to be short-lived. Conditions change and fog tends not to last forever. Fog and fire at sea are frightening scenarios and while double watches operate there will be a tense atmosphere about the navigation bridge, but it will remain the right thing to do to ensure that the vessel passes safely through any period of exceptional concern.

Ice Navigation

The threat has been around since before the days of the *Titanic*, and what a marine emergency that turned out to be! Ice in any form must be considered as dangerous, on the basis that where there is some, there is more. Just because the watch-keeper doesn't see it, it doesn't mean it isn't there.

It could be argued that in this day and age, with ships carrying high-quality radar, that an iceberg is in fact not dangerous, being visible at 20 miles. It should be noted that all icebergs remain dangerous, no matter at what range they are sighted. It is not just the berg itself, but the associated ice fragments around it, and if it is sighted at maximum radar range it would not be anticipated that the vessel would willingly navigate in near proximity to the danger.

Navigation inside known ice limits during the ice season is never a comfortable feeling. However, provided all necessary precautions are observed, a vessel could generally expect to warily proceed during the hours of daylight. In the hours of darkness, even with efficient radar, stopping the ship inside ice limits would not be considered unreasonable.

Ice is generally a poor radar target and the quality of a return echo would be dependent on the reflective surface, prevailing sea conditions and experience of the operator. The very worst example is the 'growler', which has a maximum of one metre height above the surface, worn smooth with sea water action, where radar energy is deflected in every direction but back towards the ship.

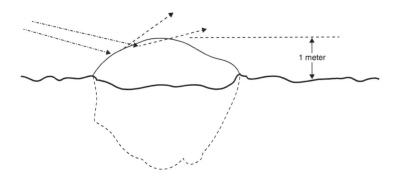

1 meter

Figure 8.1 A 'growler' showing no more than one metre above the water surface. Transmitted radar energy is deflected from the ice surface worn smooth by sea water action. The size of a typical growler equates to the same as an average-sized house. They are dangerous to shipping because they are difficult to see.

In any event, ships' Masters who encounter dangerous ice are obliged under the 1974 SOLAS to make an ice report as per the *Mariner's Handbook NP 100*.

Progress in any ice condition presents problems for the wellbeing of the vessel. In the case of 'ice accretion', if allowed to grow, the additional weight acquired could directly affect the ship's positive stability. Removing ice accretions can be an equally dangerous task for the wellbeing of crew members. While pack ice will restrict the ship's movement, the danger of becoming 'nipped' in ice formations runs the risk of the vessel being crushed. Clearly the adage of 'keep moving' even if it is an astern movement is considered better than remaining stopped.

Figure 8.2 The upper cargo deck of the *M.V. Baltic Eider*, seen proceeding through ten-tenths pack ice in the Baltic Sea. The view is taken forward through the navigation bridge windows. The *Baltic Eider* is a 1A super ice class ro-ro vessel with container carriage capacity on the upper foredeck.

Any vessel operating in ice is presented with an obstacle to overcome. It is one of those times that 'doubling of watches' might become necessary, especially if ice conditions are coupled with other restrictive conditions like fog or radar faults. Use of information from the 'ice patrol' can be beneficial and the strength of effective communications cannot be overestimated when navigating in ice conditions.

Figure 8.3 The *Star Princess*, a P&O cruise ship, operates near glacial ice formations in Alaskan waters. Several cruise operators conduct sailings from Vancouver towards Alaska, movements being through the inside passage as well as via open-water regions.

Man Overboard (MoB)

Fortunately, loss of a man overside is a rarity. However, it does occur from time to time, and because all life is of priority value, such an incident would warrant a distress alert.

The loss of a man is usually accompanied by additional circumstances, like bad weather, overside working as from stages or lack of concentration using the gangway. On some occasions, possibly where depression is a condition, the MoB scenario may be a suicide attempt.

Whatever the reason a man finds himself/herself in the water, the action to recover from the sea becomes immediate for the remaining persons on board. A bridge team would immediately take the lead role (assuming an at-sea situation) and place the navigation bridge on an alert status. The efficiency of the bridge team would be tested at this moment in carrying out four essential and simultaneous actions:

1 Adjust the helm towards the side on which the man fell.
2 Release the bridge wing lifebuoy and smoke float.
3 Sound the general alarm.
4 Have engines placed to stand-by, ready for immediate manoeuvre.

These actions need to be carried out in any order that is considered most practical.

Subsequent activities by the Officer of the Watch should ensure that the GPS coordinates provide an immediate position by activating the datum button and posting lookouts to keep line of sight on the casualty in the water.

Additional activities would also expect to include:

- The Master taking the conn of the vessel.
- Order and place a position on the chart.
- Change from automatic steering to manual.
- Post additional lookouts.
- Sound 'O' on the ship's whistle.
- Display the 'O' international code flag (by day) if other shipping present.
- Commence a vessel turning manoeuvre (various options).
- Continue the manoeuvre to complete the turn.
- Muster rescue boat crew and prepare to launch same.
- Report distress incident to nearest coast radio station.
- Advise incident position and details via an all-ships broadcast in the local area.

Should sight be lost of the casualty a 'datum known' search pattern should be plotted on the chart, i.e. sector search pattern.

Example Turning Manoeuvres

Williamson Turn

Figure 8.4 Williamson turn.

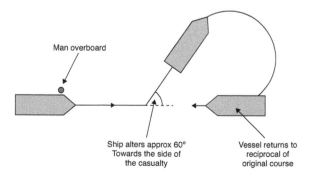

Man overboard

Ship alters approx 60°
Towards the side of
the casualty

Vessel returns to
reciprocal of
original course

Single Delayed Turn

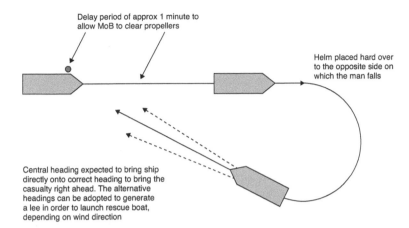

Delay period of approx 1 minute to allow MoB to clear propellers

Helm placed hard over to the opposite side on which the man falls

Central heading expected to bring ship directly onto correct heading to bring the casualty right ahead. The alternative headings can be adopted to generate a lee in order to launch rescue boat, depending on wind direction

Figure 8.5 Single delayed turn.

Double Elliptical Turn

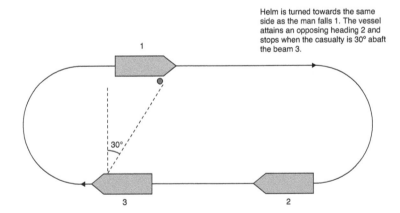

Helm is turned towards the same side as the man falls 1. The vessel attains an opposing heading 2 and stops when the casualty is 30° abaft the beam 3.

1

30°

3   2

Figure 8.6 Double elliptical turn.

Once the vessel is stopped, the rescue boat is launched to effect recovery of the MoB. The vessel continues the ellipse to a position best suited to recover the rescue craft.

The distinct advantage of this turn is that the ship's lookouts maintain line of sight on the casualty (starboard side in the example). They do not have to cross to the other side of the ship to retain line of sight as the vessel manoeuvres.

Rescue Boat Activity

To effect recovery, coxswains of rescue boats should be trained to launch in a lee created by the mother vessel. They are now also expected to be capable of launching while the mother vessel is underway and making way, up to five knots.

The approach towards the casualty in the water would be in such a manner as to recover the man on the weather side of the rescue boat, preferably in a horizontal position as opposed to a vertical lift from the water.

Boarding or Disembarking Marine Pilots

Marine pilots are regularly employed by ships moving in and around coastal waters and, by the very nature of their task, are frequently exposed to higher levels of risk of becoming the casualty in the water. Although many marine pilots are now flown to ships and board by helicopter transport, the vast majority board through fast launch pilot boats onto boarding ladders deployed by the ship. It is expected that a responsible officer would be on station at the ladder position prior to the marine pilot transferring from boat to ship. The prime function of this officer is to ensure that the pilot ladder is correctly rigged with stanchions and manropes secured. The boarding station should be well illuminated at night, a lifebuoy should be on hand as well as a heaving line.

In the event the marine pilot falls from the boat or ladder, the pilot boat itself immediately becomes the best form of rescue boat. With this in mind, coxswains of pilot boats are expected to wait in clear water until the pilot is seen to have landed safely on board the vessel before breaking away to return.

Figure 8.7 A typical pilot boat. These are usually fitted with internal rails to the centre of the deck to support pilots' movements during transfer. The boats have a high speed capacity and generally a low freeboard.

Boarding/disembarking pilots can quickly turn from routine to an emergency situation. The vertical climb on a pilot ladder should not exceed nine metres, and the age and physical fitness of the pilot are critical factors in preventing a MoB situation developing.

Navigational Pitfalls of ECDIS

The introduction of ECDIS (Electronic Chart Display and Information System) has probably been the greatest innovation to affect the shipping industry since sailing vessels gave way to steam ships. The technology change from paper charts to electronic charts is without doubt the most dynamic influence to the maritime environment in the last decade.

The system operates under IMO's performance standards, requiring vector charts and the latest information authorised by the Hydrographic Office. As with any new system, the introduction of ECDIS brought with it its own set of associated problems.

A major point of concern was that manufacturers have not standardised their operational equipment, with individual companies incorporating their own nuances within displays and operational hardware.

Training was slow to focus on the needs of operators and has becoming an additional concern for ships' Masters taking on junior Watch Officers. Adequate experience gained on designated shipboard equipment cannot always be guaranteed, especially so when Masters themselves may sometimes be unfamiliar with the modern concepts of the digital age. It is now a requirement that Masters and Navigational Officers complete both generic and ship-specific equipment ECDIS training on any ship which has ECDIS as its primary means of navigation.

Unfortunately, new technology incorporates new potential hazards for ship operators. Senior officers have variable abilities to manipulate 'passage plans' to incorporate safety features required for prudent operations. There are many benefits to the use of electronic charts, but equally they encompass many dangers. Manufacturers' display units illustrating options in different ways does not help overall confidence in active systems.

Several incidents of ships grounding have raised questions of safety settings in the electronic chart, raising concern about ECDIS usage, e.g. the *LT Cortesia* (2008) and *CSL Thames* (2011).

In order to generate a safety culture with the use of electronic chart displays, recommended safety settings have been devised to include:

- The *safety depth*: normally taken as: the ship's draft + an allowance for squat.

- Use of a *Safety Contour* established by: ship's draft + an allowance for squat + a safety margin – height of tide.
- Display changes for isolated dangers or underwater obstructions in accordance with the safety contour.

Although the industry is still in the early days of transition from paper charts to electronic chart display, one or two adversities have become apparent. Watch Officers have in many cases embraced the digital age and the electronic display, but at the same time have allowed complacency to creep into their watch-keeping practice. This has become a real danger where over-reliance on the electronic systems has overtaken the basic need to maintain an effective lookout through the bridge windows.

ECDIS is a valuable aid to the ship's navigation and sits comfortably within the integrated bridge of modern ships' designs. But it is not a standalone system. The use of visual bearings, for example, has become a more complex task for the navigator. Additionally, the many instruments that are interfaced with the electronic chart system – depth recorder, GPS/DGPS, speed log, gyro compass, ARPA and radar applications – extend the observation and monitoring skills of the Navigation Officer.

The work load at a central monitoring station can be an asset, but it can also detract the Officer of the Watch's lookout capabilities from what were previously separated work stations. The need for computer literacy has also become a necessity in many walks of life, not least in the working life of the modern-day Navigation Officer. IT work requires concentration and the real hidden danger is that ECDIS may override the other associated duties of the watch-keeper.

The use of ECDIS requires a new and revised set of skills. It is a clear advantage in the task of chart correcting, making the work simpler and easier for the navigator. The electronic chart may scroll endlessly forward, eliminating the need to change charts, but the basic navigation skills are being eroded. The built-in redundancy with two independent sources of power and the doubling up of hardware and software make ECDIS the navigation system for the future. Though the uncomfortable thought remains, what if ECDIS is lost, by whatever means – would our future navigators have retained the abilities and skills to navigate without the electrical systems in place?

Search Patterns Associated with IAMSAR

Where a MoB incident occurs it is hoped that the lookouts from the ship will not lose sight of the casualty, but this cannot be guaranteed. Where visual contact is lost the ship's Master is legally obliged to carry out a search. How long this search period would be expected to last would be at the discretion of the Master, but would be influenced considerably by the sea temperature and the sea conditions, daylight or night search and the clothing worn by the casualty when entering the water.

The recommended search pattern by IAMSAR, where the datum is known, is either a 'sector search' or an expanding square. In the case of a MoB incident the sector search is considered a suitable choice. The pattern is based over the last known position (datum), and because of the high risk of hypothermia in this case the determined track space should be considered in time rather than mileage.

Ten-minute tracks are more likely to cover the ground reasonably quickly at an operating speed of about five knots and provide a reasonable chance of recovering the casualty alive.

NB. If the casualty has contact with a lifebuoy or lifejacket, the chance of successful recovery is considerably enhanced within the operation of this pattern.

Figure 8.8 Sector search pattern.

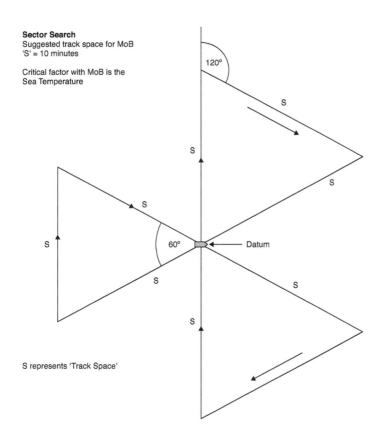

Sector Search
Suggested track space for MoB 'S' = 10 minutes

Critical factor with MoB is the Sea Temperature

S represents 'Track Space'

Figure 8.9 Coordinated creeping line surface search pattern.

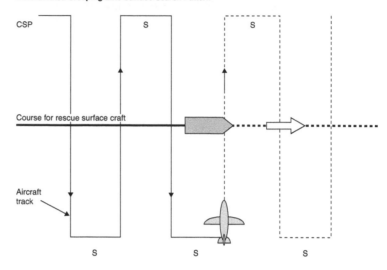

Co-ordinated Creeping Line Surface Search Pattern

S represents Track Space.

CSP represents Commencement of Search Pattern

In a designated long-range SAR operation for a variety of marine targets the most probable search pattern would be the coordinated creeping line search pattern because of the availability of aircraft of either fixed wing or rotary wing transports. The hover ability of the helicopter is beneficial, but its activity is limited by fuel range. The fixed wing (other than seaplane) cannot effect recovery of the casualty, it can only locate and engage communications to draw in a recovery surface craft.

The target definition, state of visibility, height of eye and sea conditions would all influence the recommended track space for operation of the search.

Determination of Track Space

Any potential search and rescue (SAR) operation will probably be associated with a search pattern of a type suitable to meet the circumstances of the incident. Where surface craft are involved in a designated pattern, the Master of each vessel will determine the most appropriate track space in order to keep his ship safe and bring a successful conclusion to the operation. Choice of track space is usually set as a specified range, but may be established in 'time intervals' as with an MoB/sector search.

To establish any track space the Master would inevitably consider any or all of the following:

- target definition (visible size if any above the surface);
- daylight or night conditions prevailing;
- visibility state prevailing (rain, fog, mist, snow effects);
- quality of radar target presented by target;
- sea state (number of white horses present causing obscurity);
- number of search units employed in the operation;
- height of eye of lookouts and maximum range of visible horizon;
- speed of search unit engaged in operation;
- Master's experience;
- intended search area and time factor before nightfall;
- recommendations from Marine Rescue Co-ordination Centre (MRCC) may be influential on Master's choice;
- height above sea level where aircraft are involved.

Additional factors to consider are:

- length of search may be restricted due to the endurance of the search unit;
- night searches may be ongoing with search lights, but may need to operate at reduced track space;
- the target may be able to make itself more prominent with self-help capability by means of pyrotechnics, flares, rockets, light signals, etc.

Duties of the On-Scene Coordinator

The essential function of the On-Scene Coordinator (or On Scene Commander in the military) (OSC) is to act as a communication platform between the MRCC and individual search units to bring about a successful location and recovery of the casualty.

In order to achieve this outcome, any Master who accepts the role of OSC should:

- Obtain maximum information on the target and plot a 'datum' (sources: MRCC, target itself, witnesses to last known position).
- Establish position and status of all search unit(s) in relation to datum position (course, speed, endurance, ETA, rescue facilities and medical capabilities).
- Advise all search units on search area coordinates and designated areas of operation. Recommend search pattern, track space and commencement of search pattern (CSP).
- Maintain a detailed running log account of communications and events (times, dates, places, names, results and outcomes).

- Plot all search units and their endurance.
- Request resources from MRCC, AMVER, Coast Guards via coast radio station (helicopter assistance, survival equipment, etc.).
- Debrief survivors and amend search patterns (update MRCC).
- Allocate and guard radio-specific channels/frequencies.
- Obtain and dispense weather reports to all search participants.
- Direct and coordinate activities to recover and return survivors to a safe haven.

It must be anticipated that the OSC is operating with a full bridge team in situ and that the vessel is on a full alert status. Regular communications would be expected to include progress reports and any changes to target definition, if known.

Example Checklists

Preparations by Vessel Proceeding to Vessel/Aircraft Distress Site

Table 8.1 Shipboard management table for rescue vessel on route to destress situation

Navigation Aspects	Seamanship Aspects
Plot rendezvous position	Turn out rescue boat(s) for launch
Plot own position and course	Prepare hospital for survivors
Engines to best possible speed	Note deviation in log book
Establish bridge team in position	Post lookouts in high positions
Radar long-range scanning	Change to manual steering
Highlight navigation dangers	Rig 'guest warp'
Obtain ETA at distress scene	Turn out gangway or ladders
Obtain weather forecast	Prepare rescue boat crew
Estimate currents and drift	Brief operational personnel
Obtain time of sunset	Prepare pyrotechnics and LSA gear

Communication Checklist

En route towards distress:

Table 8.2 Rescue vessel communications

Internal and External Communications
Stand-by communications officer
Acknowledge distress signal and confirm own position – responding
Establish internal communications – deck, engine room, bridge
Advise owners/company/agents to re-schedule ETA to destination
Update MRCC via OSC – ETA and own capabilities
Update target information if applicable
Obtain weather information for immediate area
Contact AMVER or other ship reporting system

- Complete radio/communication log book.
- Complete deck and engine room log books through period of deviation.
- Circumstances may require a search pattern to be plotted prior to arrival at the CSP position; in such an event, increased communication could be expected between search unit(s), OSC and MRCC, i.e. search speed, area coordinates, track space employed, visibility conditions, etc.

The Activities of the US Coast Guard

The US Coast Guard continues to conduct multiple activities across the marine environment, inclusive of operating:

- The Ice Patrol service;
- SAR operations and exercise of pollution control;
- ship reporting (AMVER), main offices in New York and San Francisco;
- Anti-smuggling patrols and drug interdiction into and around the coastline of the US.

They also conduct vessel safety inspections in and around US ports.

> **NB.** The North Atlantic Ice Patrol Service operates between 15 February and 1 July each year.

Various means of transport are employed by the organisation by way of armed surface craft, helicopters and fixed winged aircraft. They tend to work alongside and in conjunction with the US military forces towards the safety of life at sea and protection of the marine environment.

Figure 8.10 A Hercules HC-130 on low altitude patrol with the US Coast Guard.

Emergency Communications

Via NAVTEX

The NAVTEX system has become extremely popular over the last 20 years. It is used for the automatic broadcast of localised maritime safety information (MSI) employing radio telex, also known as narrow band direct printing. The system is designed for use in GMDSS Sea Area A, and operates on the medium frequency band just below the AM broadcast band. Its operational range is about 300 nm from the transmitting station.

Broadcast frequencies are as follows:

- 518 kHz is the main NAVTEX channel.
- 490 kHz is used for broadcasts in local languages (non-English).
- 4209.5 kHz is allocated for NAVTEX broadcasts in tropical areas.

The main code messages of the system are:

- Message 'A': navigational warnings originating from the Hydrographic Office (local and pilot information is not included).
- Message 'B': meteorological warnings and typical gale warnings, originating from the Meteorological Office.
- Message 'C': Ice reports, unlikely to affect the United Kingdom.
- Message 'D': SAR information; first details only affecting a casualty. Subsequent information through coast radio station on the distress frequency.
- Message 'E': meteorological forecasts.

The above message formats are frequently associated with emergency incidents.

Emergency Shoreside Communications

Whenever a maritime incident occurs, communications – both internal and external – are inevitably involved. Following preliminary investigation and possibly internal damage reports, time could generally lead to URGENCY or a PRIORITY ONE distress call.

At a point somewhere between the first concerns and the need to despatch a distress signal, the need for shoreside support should become evident. This realisation can be expected to involve contact with the designated person ashore (DPA).

Since the International Safety Management (ISM) system evolved, ship managers have a designated role ashore to provide back-up and support service to vessels which find themselves involved in situations away from their home port. Agents and services are likely to become immediate requirements to support ships in need in foreign waters. Legal support is also very often at the forefront of many incidents such as collision, pollution, territorial incursion or personnel injury. Legal representation for ships' Masters and crews could become of immediate concern.

Sensitive communications associated with any maritime incident in today's high-technological world have become a security risk, with criminal and terrorist elements capable of intercepting priority contacts. Mobile phones are now a recognisable feature among Somalian pirates and other organised criminal factions. In order to avoid interception of livefeed transmissions, the use of incident codes could be applied. This would reduce and possibly eliminate the need to resort to plain-language communications. Breaches in security could be avoided by means of pre-agreed codes, assuming such codes have been pre-set between mobile vessel units and the base shipping companies representative DPA.

Alternative message security can also be achieved through consul offices by use of diplomatic channels, but this system cannot always guarantee that a diplomatic consul is available in the immediate area. Secure communications are not easy to achieve in any event. Masters may unfortunately have weak links within their own crews or passengers. This is especially so where leaks to media outlets take place when payments for sensational stories are made available. It is inconceivable in this day and age that a cruise liner with over 2,000 people on board would not have any mobile telephones.

It is suggested that, unless security can be achieved following an incident, a prepared press statement should be issued directly to the media. This way a defined report can be despatched via the DPA in a controlled manner. It avoids speculation and eliminates a platform for difficult questions. The geographic position may also make further investigation difficult or impossible. The vessel would be initially controlling the flow of limited information rather than allowing distorted impressions to be made.

The Use of Distress Signals

Many of the incidents described by this text have led to the use of distress signals in the past. Future incidents are also more than likely to have cause to generate a signal in one form or another, to signify a distress situation. It is therefore relevant to include this communication topic within this work.

Many signals are routinely employed within the maritime environment. The majority are designated to a specific incident or operation without raising the profile to a distress scenario. A typical example can be seen from the three black spheres exhibited by the grounded vessel as its day signal. This is not a signal for help – the ship so aground does not require the assistance of other vessels. An approach by another vessel could in fact double the problem by causing the observing ship to run aground also.

The signal is specifically to indicate a ship aground, not a distress situation. Distress signals are transmitted to indicate a cry for help and assistance. By law they can only be used on the authority of the Master or Officer in Charge of a vessel, when his ship is in serious and imminent danger and there is an obvious threat to life.

It should be realised that Coast Guard authorities would rather know sooner than later of any potential life-threatening incident.

Once an appropriate authorised distress signal has been made, it is considered as a PRIORITY ONE signal, but as circumstances change this can be downgraded to an URGENCY signal, priority '2'. The obligation of the Master to cancel and revoke the distress signal once the danger has passed is a statutory requirement.

References for distress signals and their use can be found in the following publications:

- Annex 4 of the *International Regulations for the Prevention of Collision at Sea.*
- *The International Code of Signals*, statutory publication carried by all ships.
- *International Aeronautical Maritime Search and Rescue* (manual).
- The *Annual Summary of Admiralty Notices to Mariners NP 247.*
- The *Mariner's Handbook NP 100.*
- *SOLAS* (consolidated edition 2009) Chapter V, Safety of Navigation, Regulations 33–35.

Internationally Recognised Distress Signals

- The word MAYDAY spoken three times as a message by radio over VHF channel 16 (156.8 MHz) and/or high frequency (longer-range HF) on 2182 MHz.
- Transmission of a digital distress signal by activation of the distress button (or key) on a marine radio equipped with digital selective calling (DSC) over the VHF (Channel 70) and/or HF bands.
- Transmission of a digital distress signal by activation of the distress button (or key) on an Inmarsat-C satellite internet device.
- Transmission of the Morse code group *SOS* by any means.
- Emitting a volume of orange-coloured smoke.
- Raising and lowering slowly and repeatedly both arms out-stretched to each side.
- Making a continuous sounding with any fog signalling apparatus. (It may be better to use the more distinctive *SOS* with the apparatus.)
- The firing of a gun or other explosive signal at intervals of about one minute.
- Displaying the international code flags of 'NC' – November Charlie.
- Displaying a visual signal consisting of a square flag hanging above or below it a ball or anything resembling a ball (two distinctive shapes known as the distance signal).
- Launching or firing red distress rockets, throwing red stars.
- Showing flames on the vessel as from a burning oil/tar barrel.
- Burning of a red hand flare or parachute rocket.

Additionally, distress signals can be generated using automated radio signals as from a search and rescue transponder (SART) which responds to 9 GHz radar signal, or an emergency position-indicating radio beacon (EPIRB) which operates on the 406 MHz radio frequency. EPIRB signals are received and processed by a constellation of satellites of the COSPAS-SARSAT system.

A GPIRB is an EPIRB that is fitted with a GPS sensor which enables the distress signal to incorporate the position of the party in distress in latitude and longitude.

North American SAR agencies also recognise a sea dye marker and a high intensity white strobe light flashing 60 times per minute.

Annex 1

Question and Suggested Answers for Senior Officers: Towards Marine Examinations

Running Aground/Beaching

Q1. After the vessel has suddenly run aground, what would be the immediate actions expected of the ships Chief Officer?

Ans. The C/O would be expected to carry out an initial damage assessment and:

1. investigate the watertight integrity of the hull.
2. investigate whether the engine room was wet or dry.
3. make a casualty report.
4. check for any occurrence of pollution.

Q2. After making an initial damage assessment to the ship's Master, what subsequent actions would a ship's Chief Officer carry out after a grounding incident?

Ans. The Chief Officer would be expected to make a more detailed assessment of the vessel's condition by carrying out a full set of internal tank soundings and a set of external soundings. The tank soundings would provide indication of any broached tanks, while the external soundings would indicate the nature of the sea bed that the vessel is aground on and show if the vessel had adequate depth to turn propellers. Additionally, he would be expected to walk back both anchors to ensure the vessel does not accidentally float off on a rising tide before the Master is ready to attempt to refloat. It must also be considered a prudent action to seal the upper deck to ensure that any oil rising from a broached oil tank would be prevented from going overside.

NB. Water pressure in broached fuel tanks would cause oil content to rise up air and sounding pipes. Blocking off scuppers on the upper deck could prevent a pollution incident. In the event of hull damage and water ingress the Chief Officer would be expected to re-calculate the ship's stability, employing the damage stability data.

Q3. Following the vessel running aground, what would be the expected action of the ship's Master, assuming the Officer of the Watch has stopped engines?

Ans. As with any on-board maritime incident, the ship's Master would expect to take the conn of the vessel in the position of the navigation bridge. He would immediately order the closing of all watertight doors. Having ordered the ship's Chief Officer to obtain an initial damage assessment, the Master would place a position on the chart and order the Navigation Officer to establish the next times and heights of high water / low water.

He would order the Communications Officer to obtain a local weather forecast as soon as possible and for him to remain on stand-by for transmitting emergency communications.

NB. The main duty of the ship's Master in such an incident would be to engage in external communications, but he would need to establish the ship's position and obtain the initial damage assessment before opening lines of communication.

Q4. What structural members would the Master order his Chief Officer to look at in his damage assessment in the event of the ship running aground?

Ans. The Master would expect a report on the collision bulkhead and the condition of the 'tank tops'. Provided the collision bulkhead was not cracked and no rocks had pierced the tank tops, the main body of positive stability of the vessel would probably be intact.

Q5. After a grounding incident in soft sand, where the hull is seen to be free of damage, the Master decides to attempt to refloat the vessel. What must he/she have in place before any attempt is made to refloat the vessel?

Ans. Any Master would want to try and refloat once a damage assessment has revealed no damage to the hull. However, refloating must be carried out safely and not risk incurring additional damage in the process of refloating. To this end, the prudent Master would have a 'stand-by' vessel in attendance prior to attempting a refloat operation. The stand-by vessel would be expected to have the capacity to accept all persons from the grounded vessel in the event the vessel incurs excessive damage when moving towards deep water and where evacuation of crew becomes a necessity.

Q6. Following a grounding incident, soundings reveal water inside the fore end. How would the Master know whether the water is from a hull broach or from damage to the fresh water fore peak tank?

Ans. Order the damage control party to taste the water inside the hull. If it was fresh water it could be from the fore peak tank. If it was salt water it would be from outside (assuming that the fore peak is employed as a fresh water tank, not a ballast tank).

Q7. Why is it necessary to take external soundings overside after a vessel has run aground?

Ans. The ship's Master would wish to know if the depth of water around the propeller was adequate to permit the propeller(s) to turn. Second, by arming the lead the nature and composition of the sea bed would corroborate his position on the chart.

Q8. Why would a ship's Master want to deliberately beach his vessel?

Ans. A Master's actions to beach a ship would inevitably be a necessary action to save the vessel from becoming a total constructive loss. Some smaller craft like fishing boats sometimes beach in soft sand on a falling tide to expose the keel and lower hull to allow 'careening' – cleaning of the hull.

Q9. When aground, the regulations for the prevention of collision at sea recommend that she may make an appropriate signal to an approaching vessel to warn of shoals in the vicinity. What signal is considered appropriate?

Ans. It is recommended that the international code of signals of either 'U' or 'L' are given, meaning: 'You are running into danger' or 'You should stop your vessel instantly'.

Q10. If your vessel was approaching a known vessel aground, would you consider an offer of help to the grounded vessel?

Ans. No. If the grounded vessel wanted assistance I would expect her to make a distress signal. No signal of distress would indicate that she doesn't require help. Any move towards a vessel aground may only cause one's own vessel to be compromised and assistance would not be expected without a distress signal being exhibited.

Abandonment

Q1. How is the order to 'abandon ship' given and by whom?

Ans. The order to abandon ship is given by word of mouth, by the Master or the Officer in Charge.

Q2. When about to take to the boats in an abandonment situation, what additional items would be considered appropriate to add to the lifeboat/liferaft resources?

Ans. Basic survival equipment and rations are included in standard equipment for survival craft, but additional useful items include: SART, EPIRB, walkie talkie radios, blankets, extra water containers, medical supplies, a pack of playing cards, additional fuel, bridge distress rockets, transistor radio, binoculars, any spare food.

Q3. If in charge of a liferaft, what procedure would you adopt in order to beach the raft?

Ans. Look for a sandy foreshore and:
1 Man the paddles with open access points.
2 Inflate the double floor of the raft.
3 Order all personnel to don lifejackets.
4 Stream the drogue to slow down the approach.

Q4. After landing on a beach following abandonment, what would you do with the survival craft (boat or liferaft)?

Ans. It would be prudent to drag the survival craft up the beach clear of the surf. The survival craft carries all the immediate life-support requirements needed by distressed persons. The craft itself would make a bigger target for location from the air and it may need to be reused as a means of transport. In some cases improvised use of the survival craft could provide a degree of shelter from inclement weather.

Q5. A 25-man sized liferaft is in good condition and full with 25 persons. An additional 12 would-be survivors are still in the water. Could the liferaft pick up the additional 12?

Ans. The buoyancy of the liferaft has the capability to support double the capacity of the raft, so technically it could support as many as 50 persons in total. However, the floor area nor the food rations could support a 100 per cent overload. Full use could be made of the becketted lifeline on the outside of the raft, so by rotating people in and out of the raft the additional 12 could be supported, but on reduced rations.

Q6. Following an abandonment, three liferafts come together. The occupants of two of the rafts are passengers with one senior officer. The third liferaft contains four passengers, all deceased. What action would the senior officer take?

Ans. The officer should initially join the rafts together and spread the survivors evenly between two of the rafts. He should order the bodies of the deceased stripped of everything except underwear. The liferaft should be cleared of all food, water and associated equipment and resources pooled between the living. The bodies of the deceased should be left alone in the raft and the access doorways closed up. All three liferafts should then be joined together at about 30 metres distance apart, if possible. This distance would reduce the effects of snatching and parting the lines in a choppy sea. The additional liferafts would make a larger target and make the rafts more visible from the air.

NB. Close inspection of the bodies should be made to make sure life has expired. Symptoms of acute hypothermia can often be mistaken for death. Once life has expired it should be realised that the dead cannot make use of anything, but the living can utilise everything.

Q7. A large passenger vessel is involved in an abandonment. How can the passengers be managed and controlled from the public room, muster points towards the survival craft and the embarkation decks?

Ans. All crew members who deal with passengers are expected to have crowd control training and as such should be capable of managing groups of people. Group size is a critical factor and it is estimated that groups up to about 25 persons can be reasonably effectively controlled. With this in mind, keeping families together whenever possible would make sense. It is suggested that a group of 25 could be crocodiled (hand on the shoulder of the person in front to follow the leader) and led towards the survival craft, being led by a crew member and trailed by a certificated lifeboat man.

Such a method would provide an experienced lead man to deliver the group to the person in charge of the boat. The number 25 is frequently a multiple of the boat capacity, i.e. 25-man liferafts, 50-, 75- or 100-man capacity boats. Once boarded with names listed, survival craft can be launched and cleared when full, as per the muster lists.

NB. Even with the best of training, situations can always still go out of control and panic among crowds may make absolute control impossible. A well-trained crew will need to show power of command and assertiveness, as well as common sense when asked to control large groups of anxious and concerned people. A confident attitude by crew members will instil confidence within the groups they manage.

Q8. What would be considered one of the prime directives for crew members when abandoning passengers from a large passenger vessel?

Ans. Ensure that all passengers are wearing correctly donned lifejackets and warm clothing.

Q9. A passenger vessel must carry a rescue boat on either side of the vessel. What are the two functions of the rescue boats when involved in an abandonment situation ?

Ans. The rescue boats are expected to recover persons from the water and marshal survival craft together.

NB. Liferafts are not motorised and as such are difficult to manoeuvre. The rescue boats are equipped with a 50 metre towing line to effectively draw liferafts away from dangerous situations towards positions of relative safety.

Q10. A chemical tanker is on fire and being abandoned. Survival craft are about to be launched; what instructions would the Master give to the coxswains of lifeboats?

Ans. The increased possibility of toxic substances being released from this type of vessel are high. As such, Masters would order coxswains to head their boats upwind to ensure dangerous substances are carried downwind. Additionally, it would be expected that survival craft stay together, at about two miles from the distress position unless there is an immediate safe haven.

Fire on Board

Q1. What are the two securities of the CO_2 total flood system against accidental release of the gas?

Ans. Close to the locked remote cabinet for CO_2 activation is a small key holder where a break-glass-to-obtain-key setup is positioned. Second, a gate stop valve is set in the line that is only opened once the pilot bottles are fired.

Q2. What is the danger of directing water onto a cargo hold full of bulk coal when on fire?

Ans. Bulk coal fires are generally caused by methane gas being released from the cargo. Coal fires are known to be very hot and water use with its good knockdown cooling capability would seem the obvious choice of firefighting medium. However, water should not be used as this will turn to steam on contact with the hot surface. This steam will in turn pressurise the hold and could cause the hatch tops or the ships sides to be blown out.

A suggested resolution is to keep the cargo hatch battened down and commence boundary cooling, so starving the fire of oxygen. The ship should alter course towards a suitable port of refuge where the hatches can be opened. Water can then be used to extinguish the fire and the steam can be escaped to the atmosphere. The fire brigade ashore could be utilised and cranes with grabs could dig out the fire within the bulk cargo. Flooding the hold may also be a suitable option, provided the stability of the vessel allows this.

Q3. What precautions would be expected when placing crew members into fire suits and breathing apparatus prior to fighting a fire?

Ans. Assuming the outfits have no defects, it would be expected to check that the air bottle is full by inspection of the gauge and making sure the indicator is in the green section. Once the air is turned on, a warning whistle should sound. The mask should be donned by the wearer and a gas seal around the airways should be established. This could be checked by briefly shutting off the air supply and seeing the mask 'crush' onto the wearer's face. This would indicate that the gas seal is good and smoke inhalation should not affect the wearer. The life line should be connected to the back of the harness.

A monitoring board should be opened to time when firefighters enter the space compartment and when they are due to return, bearing in mind that the average air bottle provides approximately 30 minutes of air under normal working conditions.

Q4. What is the purpose of the crash panel built into the base of accommodation cabin doors ?

Ans. The crash panels are included to eliminate the need to open the cabin door in the first instance when tackling a fire. It allows a jet of water to be directed to the deckhead via the crash panel. As the water is deflected from the deckhead it is expected to cool the inside of the cabin down, allowing fire fighters to enter into a less hostile fire environment.

Q5. What are the advantages of a water mist fixed system over and above a total flood CO_2 firefighting system?

Ans. The water mist system allows persons to still breath inside the protected space. It also has good knockdown capability and a better cooling effect. CO_2 is a smothering agent, eliminating the oxygen content. People cannot breathe inside the CO_2 environment and neither will the gas have as good a cooling effect as the water system.

Q6. Helicopters present an inherent fire risk to a vessel, but when are they most vulnerable to causing a fire?

Ans. Most helicopter/ship operations engage in hoist operations rather than in a land-on exercise. As such, they may shut down their motors once landed on deck. It is when they restart engines prior to departure that the higher risk of a fuel surge may occur, causing an increased risk of fire.

NB. Offshore operational helicopters have their own built-in firefighting systems and ship's firefighters are not expected to engage with conventional firefighting apparatus unless the pilot requests their input.

Q7.   A vessel experiences a fire when secured alongside in a port. The fire alarm has been raised and the local fire brigade is en route to the ship. Whose authority will prevail throughout the incident? Will it be the authority of the Master or that of the Fire Brigade Chief?

Ans.   The ship's Master or his designated agent will be the prevailing authority.

Q8.   When attending a fire aboard a ship in port, the local Fire Brigade Officer would expect to have access to certain shipboard documentation. What documents would you expect to pass to the Fire Officer when he boards?

Ans.   The Fire Brigade Officer would expect to receive the following:
   • a cargo plan of the vessel, with list of hazardous cargoes;
   • a list of names and ranks of all persons on board;
   • a list of crew members known not to be on board;
   • a copy of the ship's firefighting arrangement plan;
   • a copy of the ship's general arrangement plan.

Q9.   If fighting a fire alongside in port, with water as a firefighting medium, when does the amount of water being used as an extinguishing agent not detrimentally affect the stability of the vessel?

Ans.   Water as a firefighting medium could detrimentally affect the positive stability of the vessel. Where the under-keel clearance of the ship is less than one-ninth of the ship's draught, the volume of water will not generate an unstable condition because mean sinkage will cause the vessel to sit on the sea bed/harbour bottom.

Q10.   What is the danger associated with tackling a cargo fire with hoses where the cargo is known to have a high dust concentration?

Ans.   Dusty cargoes like grain cargoes could experience a static charge from the hose impact, causing a 'dust explosion'. Such an explosion can be expected to be violent and volatile, causing damage or injury to personnel in the vicinity.

Miscellaneous Hazards

Q1.   What are the predominant means and methods of communication around the UK coastline if and when involved in an SAR operation?

Ans.   Communications between Coast Guard rescue teams, inshore lifeboats (RNLI) and other nominated SAR assets – both MCA and MoD – typically take place over VHF marine radio, MF radio and telephone (Satellite, landline or mobile).

Q2. When approaching a new danger marked by double cardinal marks, where would you expect to obtain updated information and how could the danger be pinpointed?

Ans. Updated information on new dangers could be found in navigation warnings, such as from NAVTEX. The danger could be identified and pinpointed because it would exhibit a RACON emitting Morse code 'D', showing a radar signature of 1 nm on radar.

Q3. What would be the requirements and capabilities for a vessel to act as the On-Scene Coordinator (OSC) if and when it is required to conduct a search operation under the guidelines of IAMSAR?

Ans. The ideal vessel for coordinating a search pattern would be a warship, preferably one with air support, acting as On Scene Commander. In the absence of such a vessel, a suitable commercial vessel with the following attributes would be expected to coordinate additional search units:
- effective all-round short- and long-range communications;
- adequate fuel and store reserves;
- sufficient manpower to fulfil all respective duties;
- adequate plotting facilities;
- no commercial pressure from sensitive cargoes;
- suitable position from the incident position;
- Master with search experience preferred;
- reliable machinery and capable endurance.

NB. Actual rescue and recover capability is a requirement of the search units and the OSC would not require such rescue facilities unless participating in a dual role of OSC/SU.

Q4. When acting in the role of a search unit (SU) engaged in an IAMSAR search pattern, what factors would determine the 'track space' you would adopt to complete the search pattern?

Ans. The designated track space for conducting a search pattern may be recommended by the OSC or the mission coordinator ashore at MRCC. However, the Master of the search unit actually conducting the search would determine what track space is practical. The factors influencing track space would be:
- target definition (size above surface and colour);
- state of visibility prevailing and visible range;
- quality of radar target likely to be presented;
- night or daylight search (night search ongoing with search lights);

- number of search units involved and their proximity;
- height of eye of lookouts and visible range;
- speed of vessel in ongoing search;
- Master's experience;
- sea state and the number of 'white horses';
- search area limits and endurance of the searching vessel;
- time factor in relation to period of remaining daylight;
- recommendations from MRCC;
- any known self-help capability from target itself to use pyrotechnics;
- recommended track spaces for targets from IAMSAR manual.

Q5. How would a helicopter wish to engage in a winching operation with a ship on the vessel's port side?

Ans. The ship would be expected to make a course with the wind about 30° on the port bow. The offshore helicopter would have its winch and access on its starboard side. As such the engagement would be conducted starboard side of the aircraft to port side of the ship, with both heading into the wind.

Q6. Following a 'man overboard' incident, a Master manoeuvres his vessel and turns the ship in open water. The ship returns to the last known position of the casualty but lookouts report no sighting of the casualty. What must the Master now do?

Ans. The Master is legally bound to carry out a search of the area and make a distress call, giving the position of the incident and associated details. The search would probably be a 'sector search' pattern as recommended by IAMSAR, where the datum is known. How long the Master decides to keep the search going is at his discretion and would probably depend on several factors, especially sea temperature and prevailing weather/sea conditions, together with what period of daylight is left. Results of the search must be communicated and the distress cancelled. A statement must be entered into the Official Log Book (OLB) to reflect the incident, with the results of the search.

Q7. Passenger vessels are required to carry an on board search and rescue plan. What is contained in the ship(s) element of the plan?

Ans. The SAR plan would contain the general particulars of the ship, including the MMSI number, call sign, country of registration, type of vessel, colour scheme, gross tonnage, overall length, maximum draught, service speed, maximum number of persons allowed on board and the number of crew.

Also, additionally profile plans and deck plans highlighting all lifesaving facilities, firefighting equipment and helicopter landing/winch areas, with respective approach sectors, and the type of aircraft for which the construction is acceptable.

The plan would also cover specific methods that could be employed to recover persons from the sea or from another vessel. It would also show a list of the communication equipment carried by the ship, and indicate whether the above details could be transmitted by electronic means.

Q8. When approaching an active search and rescue transponder (SART), what is the expected range and radar signature that would be expected on the search radar of a surface vessel?

Ans. A SART could expect to be detected at about five nautical miles where the active SART is at a position of about one metre above surface level. The radar signature would be presented by a series of 12 dots in a line on screen. This pattern could be expected to change to 12 arcs as the range decreases and then turn to concentric circles when the search unit is metres away.

Q9. What are the meteorological hazards that require ships' Masters to make a statutory danger message report?

Ans. Masters must report on any of the following items that are encountered:
- dangerous ice;
- storm force 10 winds for which no storm warning has been received;
- tropical cyclone that has been encountered or is developing;
- sub-freezing air temperatures associated with gale force winds.

Q10. Prior to proceeding to sea, the Master must ensure that the intended voyage has been planned using relevant nautical charts and publications. What must the voyage plan identify on the planned route?

Ans. The voyage plan must take account of:
- relevant ships' routing systems;
- sufficient sea room for the passage throughout the voyage;
- all known navigation hazards;
- the marine environmental protection measures, and avoid as far as possible any actions and activities which could damage the environment.

Annex 2

Notable Shipping Incidents

Ship Name	Ship Type	Date and Nature of Incident
Roystan Grange/Tien Chee	Reefer/tanker	May 1972, collision
Herald of Free Enterprise	Ro-ro ferry	March 1987, capsize
Exxon Valdez	Tanker	March 1989, grounding
Braer	Tanker	January 1993, lee shore
Sally Albatross	Passenger cruise	March 1994, grounding
Estonia	Ro-pax ferry	September 1994, capsize/sank
Sea Empress	Tanker	February 1996, grounding
M.V. Tricolour	Car carrier	December 2002, capsize
Cast Beaver	Container	January 2006, grounding
M.V. Calypso	Passenger cruise	May 2006, fire
MSC Napoli	Container	January 2007, flooding
Riverdance	Vehicle ferry	January 2008, grounding
Bunga Alpinia	Tanker	August 2010, fire
Costa Concordia	Passenger cruise	January 2012, grounding
MSC Flaminia	Container	July 2012, fire/explosion
Kullack	O/S drilling rig	December 2012, grounding

LOF 2011

Annex 3

LLOYD'S STANDARD FORM OF SALVAGE AGREEMENT

(Approved and Published by the Council of Lloyd's)

NO CURE - NO PAY

1. Name of the salvage Contractors: (referred to in this agreement as "the Contractors")	2. Property to be salved: The vessel: her cargo freight bunkers stores and any other property thereon but excluding the personal effects or baggage of passengers master or crew (referred to in this agreement as "the property")
3. Agreed place of safety:	4. Agreed currency of any arbitral award and security (if other than United States dollars)
5. Date of this agreement	6. Place of agreement
7. Is the Scopic Clause incorporated into this agreement? State alternative : Yes/No	
8. Person signing for and on behalf of the Contractors Signature:	9. Captain or other person signing for and on behalf of the property Signature:

A **Contractors' basic obligation:** The Contractors identified in Box 1 hereby agree to use their best endeavours to salve the property specified in Box 2 and to take the property to the place stated in Box 3 or to such other place as may hereafter be agreed. If no place is inserted in Box 3 and in the absence of any subsequent agreement as to the place where the property is to be taken the Contractors shall take the property to a place of safety.

B **Environmental protection:** While performing the salvage services the Contractors shall also use their best endeavours to prevent or minimise damage to the environment.

C **Scopic Clause:** Unless the word "No" in Box 7 has been deleted this agreement shall be deemed to have been made on the basis that the Scopic Clause is not incorporated and forms no part of this agreement. If the word "No" is deleted in Box 7 this shall not of itself be construed as a notice invoking the Scopic Clause within the meaning of sub-clause 2 thereof.

D **Effect of other remedies:** Subject to the provisions of the International Convention on Salvage 1989 as incorporated into English law ("the Convention") relating to special compensation and to the Scopic Clause if incorporated the Contractors services shall be rendered and accepted as salvage services upon the principle of "no cure - no pay" and any salvage remuneration to which the Contractors become entitled shall not be diminished by reason of the exception to the principle of "no cure - no pay" in the form of special compensation or remuneration payable to the Contractors under a Scopic Clause.

E **Prior services:** Any salvage services rendered by the Contractors to the property before and up to the date of this agreement shall be deemed to be covered by this agreement.

F **Duties of property owners:** Each of the owners of the property shall cooperate fully with the Contractors. In particular:

    (i)   the Contractors may make reasonable use of the vessel's machinery gear and equipment free of expense provided that the Contractors shall not unnecessarily damage abandon or sacrifice any property on board;

    (ii)   the Contractors shall be entitled to all such information as they may reasonably require relating to the vessel or the remainder of the property provided such information is relevant to the performance of the services and is capable of being provided without undue difficulty or delay;

    (iii)   the owners of the property shall co-operate fully with the Contractors in obtaining entry to the place of safety stated in Box 3 or agreed or determined in accordance with Clause A.

G **Rights of termination:** When there is no longer any reasonable prospect of a useful result leading to a salvage reward in accordance with Convention Articles 12 and/or 13 either the owners of the vessel or the Contractors shall be entitled to terminate the services hereunder by giving reasonable prior written notice to the other.

H **Deemed performance:** The Contractors' services shall be deemed to have been performed when the property is in a safe condition in the place of safety stated in Box 3 or agreed or determined in accordance with clause A. For the purpose of this provision the property shall be regarded as being in safe condition notwithstanding that the property (or part thereof) is damaged or in need of maintenance if (i) the Contractors are not obliged to remain in attendance to satisfy the requirements of any port or habour authority, governmental agency or similar authority and (ii) the continuation of skilled salvage services from the Contractors or other salvors is no longer necessary to avoid the property becoming lost or significantly further damaged or delayed.

I **Arbitration and the LSSA Clauses:** The Contractors' remuneration and/or special compensation shall be determined by arbitration in London in the manner prescribed by Lloyd's Standard Salvage and Arbitration Clauses ("the LSSA Clauses") and Lloyd's Procedural Rules in force at the date of this agreement. The provisions of the said LSSA Clauses and Lloyd's Procedural Rules are deemed to be incorporated in this agreement and form an integral part hereof. Any other difference arising out of this agreement or the operations hereunder shall be referred to arbitration in the same way.

J **Governing law:** This agreement and any arbitration hereunder shall be governed by English law.

K **Scope of authority:** The Master or other person signing this agreement on behalf of the property identified in Box 2 enters into this agreement as agent for the respective owners thereof and binds each (but not the one for the other or himself personally) to the due performance thereof.

L **Inducements prohibited:** No person signing this agreement or any party on whose behalf it is signed shall at any time or in any manner whatsoever offer provide make give or promise to provide or demand or take any form of inducement for entering into this agreement.

IMPORTANT NOTICES

1 **Salvage security.** As soon as possible the owners of the vessel should notify the owners of other property on board that this agreement has been made. If the Contractors are successful the owners of such property should note that it will become necessary to provide the Contractors with salvage security promptly in accordance with Clause 4 of the LSSA Clauses referred to in Clause I. The provision of General Average security does not relieve the salved interests of their separate obligation to provide salvage security to the Contractors.

2 **Incorporated provisions.** Copies of the applicable Scopic Clause, the LSSA Clauses and Lloyd's Procedural Rules in force at the date of this agreement may be obtained from (i) the Contractors or (ii) the Salvage Arbitration Branch at Lloyd's, One Lime Street, London EC3M 7HA.

3 **Awards.** The Council of Lloyd's is entitled to make available the Award, Appeal Award and Reasons on www.lloydsagency.com (the website) subject to the conditions set out in Clause 12 of the LSSA Clauses.

4 **Notification to Lloyd's.** The Contractors shall within 14 days of their engagement to render services under this agreement notify the Council of Lloyd's of their engagement and forward the signed agreement or a true copy thereof to the Council as soon as possible. The Council will not charge for such notification.

Tel.No. + 44(0)20 7327 5408/5407
Fax No. +44(0)20 7327 6827
E-mail: lloyds-salvage@lloyds.com
www.lloydsagency.com

15.1.08 3.12.24 13.10.26 12.4.50 10.6.53 20.12.67
23.2.72 21.5.80 5.9.90 1.1.95 1.9.2000 9.5.2011

Amendments to Lloyd's Standard Form of Salvage Agreement (LOF) and Lloyd's Standard Salvage and Arbitration (LSSA) Clauses

Following lengthy, open and constructive debate at the Lloyd's Salvage Group (LSG) meetings in 2010 and March 2011, the following amendments to LOF2000 and the LSSA Clauses have been agreed. The new LOF will be known as LOF2011.

LOF2011

Two new clauses, details of which appear below, have been added to LOF. They appear on page 2 under IMPORTANT NOTICES and are numbered 3 and 4 respectively.

Details of LOF Awards on Lloyd's web-site

3   *Awards. The Council of Lloyd's is entitled to make available the Award, Appeal Award and Reasons on www.lloydsagency.com (the website) subject to the conditions set out in Clause 12 of the LSSA Clauses.*

LOF Awards, Appeal Awards and Reasons have traditionally been confidential to the parties involved. However, the LSG was unanimous in agreement that the Arbitrator's Award (and where applicable, the Appeal Award) should be made more widely accessible. It was further agreed that such access will be via subscription to the appropriate area of Lloyd's website at www.lloydsagency.com.

This amendment to LOF is in line with other recent changes, including the new system for appointments to the LOF Panel of Arbitrators, designed to make the LOF process a more transparent and inclusive one.

The conditions governing the making available of Awards, etc, are set out in a new LSSA Clause 12, which is referred to separately below.

Details of how to apply for subscription to the website can be obtained from the Salvage Arbitration Branch (SAB) (see contact details below).

Notification of LOFs to Lloyd's

4   *Notification to Lloyd's. The Contractors shall within 14 days of their engagement to render services under this agreement notify the Council of Lloyd's of their engagement and forward the original agreement or a true copy thereof to the Council as soon as possible. The Council will not charge for such notification.*

It has always been the case that LOFs have been agreed and services successfully rendered without the matter being notified to Lloyd's. In most cases this was because the salvors and salved interests were able to reach a quick, amicable settlement and therefore did not require the services of the SAB and the LOF arbitration system.

However, it would appear that the number of these cases has increased over recent years and it has become very difficult to gauge the actual level of use of LOF. The salvors are now being asked to report all LOFs to Lloyd's within 14 days of their engagement. It is, of course, a requirement that attracts no charge from Lloyd's.

LSSA Clauses

Security for Arbitrator's and Appeal Arbitrator's Fees

The following new Clauses have been introduced to the LSSA Clauses:

> *6.6 The Arbitrator shall be entitled to satisfactory security for his reasonable fees and expenses, whether such fees and expenses have been incurred already or are reasonably anticipated. The Arbitrator shall have the power to order one or more of the parties to provide security in a sum or sums and in a form to be determined by the Arbitrator. The said power may be exercised from time to time as the Arbitrator considers appropriate.*

It happens from time-to-time that an Arbitrator is appointed to a particular matter in which the salvage security, which traditionally covers the fees and/or costs of the Arbitrator as well as Lloyd's, has either been provided direct to the salvors in a form that is not acceptable to Lloyd's or has not been provided at all.

The Arbitrators have become increasingly concerned at the level of their exposure to the potential non-payment of their fees. This clause gives them power to order the provision of security, for sum or sums determined by them, in respect of their reasonable fees and expenses.

> *10.8 The Appeal Arbitrator shall be entitled to satisfactory security for his reasonable fees and expenses, whether such fees and expenses have been incurred already or are reasonably anticipated. The Appeal Arbitrator shall have the power to order one or more of the parties to provide security in a sum or sums and in a form to be determined by the Appeal Arbitrator. The said power may be exercised from time to time as the Appeal Arbitrator considers appropriate.*

This clause gives the same powers to the Appeal Arbitrator as those invested in the first instance Arbitrators set out in Clause 6.6 above.

Details of LOF Awards on Lloyd's web-site

12 Awards

> *12.1 The Council will ordinarily make available the Award, or Appeal Award, and Reasons on www.lloydsagency.com (the website) except where the Arbitrator or Appeal Arbitrator has ordered, in response to representations by any party to the Award or Appeal Award, that there is a good reason for deferring or withholding them. Any party may make such representations to the Arbitrator provided a written notice of its intention to do so is received by the Council no later than 21 days after the date on which the Award or Appeal Award was published by the Council and the representations themselves are submitted in writing to the Arbitrator or Appeal Arbitrator within 21 days of the date of the notice of intention.*

> *12.2 Subject to any order of the Arbitrator or Appeal Arbitrator the Award, or Appeal Award, and Reasons will be made available on the website as soon as practicable after expiry of the 21 day period referred to in clause 12.1.*

> *12.3 In the event of an appeal being entered against an Award , the Award and Reasons shall not be made available on the website until either the Appeal Arbitrator has issued his Appeal Award or the Notice of Appeal is withdrawn subject always to any order being made in accordance with clause 12.1.*

As stated above (see LOF new IMPORTANT NOTICE 3) these new clauses set out the conditions governing the making available of LOF Awards, etc on Lloyd's website www.lloydsagency.com

Note that Lloyd's will make available the Award and Reasons on its website 21 days after publication of the Award unless:

(i)   An appeal has been entered against an Award or
(ii) The Arbitrator or Appeal Arbitrator has ordered, in response to representations by any party, that there is "good reason" for deferring or withholding them.

In the event of (i) above, the Award and Reasons will not be made available on the website until either the Appeal Arbitrator has issued his Appeal Award or the Notice of Appeal is withdrawn (subject always to any order referred to in (ii) above.)

<u>Container Vessel Cases</u>

Special Provisions

These Special Provisions shall apply to salved cargo insofar as it consists of laden containers.

> *13 The parties agree that any correspondence   or notices in respect of salved property which is not the subject of representation in accordance with Clause 7 of these Rules may be sent to the party or parties who have provided salvage security in respect of that property and that this shall be deemed to constitute proper notification to the owners of such property.*

> *14 Subject to the express approval of t   he Arbitrator, where an agreement is reached between the Contractors and the owners of salved cargo comprising at least 75% by value of salved cargo represented in accordance with Clause 7 of these Rules, the same agreement shall be binding on the owners of all salved cargo who were not represented at the time of the said approval.*

> *15 Subject to the express approv   al of the Arbitrator, any salved cargo with a value below an agreed figure may be omitted from the salved fund and excused from liability for salvage where the cost of including such cargo in the process is likely to be disproportionate to its liability for salvage.*

Discussions to implement the above (or similar) clauses have been taking place in the LSG forum over the past three years. These discussions originated out of concerns that the costs incurred in collecting salvage security from low-value cargo interests in cases involving container (multi-bill of lading) vessels were disproportionate to their proportion of any salvage award or settlement.

Clause 13 specifically relates to the provisions of the Arbitration Act 1986, which require notices to be given to the owners of the salved property, which in container vessel cases may number several hundred or even thousands. It allows the SAB, or the salvors, or their appointed representatives/agents to send any appropriate notices to the party (usually the cargo insurers) that has provided the salvage security.

This can significantly reduce the number of notices to be sent because, often, an insurer will have provided security for a number of their insureds.

It is often the case in container vessel cases that the salvors are able to reach an amicable settlement with the "represented" cargo interests, but are left with no option but to obtain an Award against the remaining interests, thereby incurring the costs associated with utilizing the full arbitration process. The provisions set out in Clause 14 above allow the salvors to apply to the Arbitrator to bind the unrepresented cargo to the terms of the settlement agreement where the agreement has been reached with owners of at least 75% by value of the salved cargo.

Clause 15 allows the salvors to apply to the Arbitrator to excuse any cargo below an agreed value from any liability for salvage where the cost of including it is likely to be disproportionate to its proportion of any Award or settlement.

The potential effect of these clauses (13, 14 and 15) is to reduce the cost of collecting salvage security and obtaining an Award against the unrepresented cargo.

The contact details for the SAB are as follows:

Salvage Arbitration Branch
Agency Department
Lloyd's
One Lime Street
London EC3M 7HA

Fax: +44 (0)20 7327 6827

Kevin Clarke
Tel: +44 (0)20 7327 540/8
Email: kevin.clarke@lloyds.com

Diane Bowles
Tel: +44 (0)20 7327 5407
Email: diane.bowles@lloyds.com

Website: www.lloydsagency.com

LOF2011 and the amended LSSA Clauses can now be downloaded from www.lloydsagency.com by clicking on the following link:

http://www.lloyds.com/The-Market/Tools-and-Resources/Lloyds-Agency-Department/Salvage-Arbitration-Branch/Lloyds-Open-Form-LOF

If you have any queries in relation to these changes please do not hesitate to contact members of the Salvage Arbitration Branch or me.

Karen

KAREN BIZON
Controller of Agencies
Lloyd's Agency Department
Lloyd's, One Lime Street, London EC3M 7HA
Telephone +44 (0)20 7327 5735
Mobile   +44 (0)7931661762
www.lloydsagency.com

Lloyd's Agency 1811- 2011: Celebrating two centuries of service to the global maritime industry

Annex 4

LLOYD'S STANDARD FORM OF SALVAGE AGREEMENT
(Approved and Published by the Council of Lloyd's)

LLOYD'S STANDARD SALVAGE AND ARBITRATION CLAUSES

1 **Introduction**

1.1 These clauses ("the LSSA Clauses") or any revision thereof which may be published with the approval of the Council of Lloyd's are incorporated into and form an integral part of every contract for the performance of salvage services undertaken on the terms of Lloyd's Standard Form of Salvage Agreement as published by the Council of Lloyd's and known as LOF 2011 (or its predecessor LOF 2000) ("the Agreement" which expression includes the LSSA clauses and Lloyd's Procedural Rules referred to in Clause 6).

1.2 All notices communications and other documents required to be sent to the Council of Lloyd's should be sent to:

    Salvage Arbitration Branch
    Lloyd's
    One Lime Street
    London EC3M 7HA

    Tel: +44 (0) 20 7327 5408/5407
    Fax: +44 (0) 20 7327 6827
    E-mail: lloyds-salvage@lloyds.com

2 **Overriding Objective**

In construing the Agreement or on the making of any arbitral order or award regard shall be had to the overriding purposes of the Agreement namely:

a   to seek to promote safety of life at sea and the preservation of property at sea and during the salvage operations to prevent or minimise damage to the environment;

b   to ensure that its provisions are operated in good faith and that it is read and understood to operate in a reasonably businesslike manner;

c   to encourage cooperation between the parties and with relevant authorities;

d   to ensure that the reasonable expectations of salvors and owners of salved property are met and

e   to ensure that it leads to a fair and efficient disposal of disputes between the parties whether amicably, by mediation or by arbitration within a reasonable time and at a reasonable cost.

3 **Definitions**

In the Agreement and unless there is an express provision to the contrary:

3.1 "Award" includes an interim or provisional Award and "Appeal Award" means any Award including any interim or provisional Award made by the Appeal Arbitrator appointed under clause 10.2.

3.2 "personal effects or baggage" as referred to in Box 2 of the Agreement means those which the passenger, Master and crew member have in their cabin or are otherwise in their possession, custody or control and shall include any private motor vehicle accompanying a passenger and any personal effects or baggage in or on such vehicle.

3.3 "Convention" means the International Convention on Salvage 1989 as enacted by section 224, Schedule 11 of the Merchant Shipping Act 1995 (and any amendment of either) and any term or expression in the Convention has the same meaning when used in the Agreement.

3.4 "Council" means the Council of Lloyd's

3.5 "days" means calendar days

3.6 "Owners" means the owners of the property referred to in box 2 of the Agreement

3.7 "owners of the vessel" includes the demise or bareboat charterers of that vessel.

3.8 "special compensation" refers to the compensation payable to salvors under Article 14 of the Convention.

3.9 "Scopic Clause" refers to the agreement made between (1) members of the International Salvage Union (2) the International Group of P&I Clubs and (3) certain property underwriters which first became effective on 1st August 1999 and includes any replacement or revision thereof. All references to the Scopic Clause in the Agreement shall be deemed to refer to the version of the Scopic Clause current at the date the Agreement is made.

4 Provisions as to Security, Maritime Lien and Right to Arrest

4.1 The Contractors shall immediately after the termination of the services or sooner notify the Council and where practicable the Owners of the amount for which they demand salvage security (inclusive of costs expenses and interest) from each of the respective Owners.

4.2 Where a claim is made or may be made for special compensation the owners of the vessel shall on the demand of the Contractors whenever made provide security for the Contractors' claim for special compensation provided always that such demand is made within 2 years of the date of termination of the services.

4.3 The security referred to in clauses 4.1. and 4.2. above shall be demanded and provided in the currency specified in Box 4 or in United States Dollars if no such alternative currency has been agreed.

4.4 The amount of any such security shall be reasonable in the light of the knowledge available to the Contractors at the time when the demand is made and any further facts which come to the Contractors' attention before security is provided. The arbitrator appointed under clause 5 hereof may, at any stage of the proceedings, order that the amount of security be reduced or increased as the case may be.

4.5 Unless otherwise agreed such security shall be provided (i) to the Council (ii) in a form approved by the Council and (iii) by person firms or corporations acceptable to the Council or acceptable to the Contractors. The Council shall not be responsible for the sufficiency (whether in amount or otherwise) of any security which shall be provided nor the default or insolvency of any person firm or corporation providing the same.

4.6 The owners of the vessel including their servants and agents shall use their best endeavours to ensure that none of the property salved is released until security has been provided in respect of that property in accordance with clause 4.5.

4.7 Until security has been provided as aforesaid the Contractors shall have a maritime lien on the property salved for their remuneration.

4.8 Until security has been provided the property salved shall not without the consent in writing of the Contractors (which shall not be unreasonably withheld) be removed from the place to which it has been taken by the Contractors under clause A. Where such consent is given by the Contractors on condition that they are provided with temporary security pending completion of the voyage the Contractors' maritime lien on the property salved shall remain in force to the extent necessary to enable the Contractors to compel the provision of security in accordance with clause 4.5.

4.9 The Contractors shall not arrest or detain the property salved unless:
   (i)   security is not provided within 21 days after the date of the termination of the services or
   (ii)   they have reason to believe that the removal of the property salved is contemplated contrary to clause 4.8. or
   (iii) any attempt is made to remove the property salved contrary to clause 4.8.

5 Appointment of Arbitrators

5.1 Whether or not security has been provided (and always subject to Clause 6.6 and 10.8 below) the Council shall appoint an arbitrator ("the Arbitrator") upon receipt of a written request provided that any party requesting such appointment shall if required by the Council undertake to the Council's reasonable satisfaction to pay the reasonable fees and expenses of the Council and those of the Arbitrator and the Appeal Arbitrator.

5.2 The Arbitrator, the Appeal Arbitrator and the Council may charge reasonable fees and expenses for their services whether the arbitration proceeds to a hearing or not and all such fees and expenses shall be treated as part of the costs of the arbitration.

6 Arbitration Procedure and Arbitrators Powers

6.1 The arbitration shall be conducted in accordance with the Procedural Rules approved by the Council ("Lloyd's Procedural Rules") in force at the date of the LOF agreement.

6.2 The arbitration shall take place in London unless (i) all represented parties agree to some other place for the whole or part of the arbitration and (ii) any such agreement is approved by the Arbitrator on such terms as to the payment of the Arbitrator's travel and accommodation expenses as he may see fit to impose.

6.3 The Arbitrator shall have power in his absolute discretion to include in the amount awarded to the Contractors the whole or part of any expenses reasonably incurred by the Contractors in:

    (i) ascertaining demanding and obtaining the amount of security reasonably required in accordance with clause 4.5;

    (ii) enforcing and/or protecting by insurance or otherwise or taking reasonable steps to enforce and/or protect their lien;

    (iii) securing the payment of the fees and expenses of the Council, the Arbitrator and the Appeal Arbitrator.

6.4 The Arbitrator shall have power to make but shall not be bound to make a consent award between such parties as so consent with or without full arbitral reasons

6.5 The Arbitrator shall have power to make a provisional or interim award or awards including payments on account on such terms as may be fair and just

6.6 The Arbitrator shall be entitled to satisfactory security for his reasonable fees and expenses, whether such fees and expenses have been incurred already or are reasonably anticipated. The Arbitrator shall have the power to order one or more of the parties to provide such security in a sum or sums and in a form to be determined by the Arbitrator. The said power may be exercised from time to time as the Arbitrator considers appropriate.

6.7 Awards in respect of salvage remuneration or special compensation (including payments on account) shall be made in the currency specified in Box 4 or in United States dollars if no such alternative currency has been agreed.

6.8 The Arbitrator's Award shall (subject to appeal as provided in clause 10) be final and binding on all the parties concerned whether they were represented at the arbitration or not and shall be published by the Council in London.

7 Representation of Parties

7.1 Any party to the Agreement who wishes to be heard or to adduce evidence shall appoint an agent or representative ordinarily resident in the United Kingdom to receive correspondence and notices for and on behalf of that party and shall give written notice of such appointment to the Council.

7.2 Service on such agent or representative by letter, e-mail or facsimile shall be deemed to be good service on the party which has appointed that agent or representative.

7.3 Any party who fails to appoint an agent or representative as aforesaid shall be deemed to have renounced his right to be heard or adduce evidence.

8 Interest

8.1 Unless the Arbitrator in his discretion otherwise decides the Contractors shall be entitled to interest on any sums awarded in respect of salvage remuneration or special compensation (after taking into consideration any sums already paid to the Contractors on account) from the date of termination of the services until the date on which the Award is published by the Council and at a rate to be determined by the Arbitrator.

8.2 In ordinary circumstances the Contractors' interest entitlement shall be limited to simple interest but the Arbitrator may exercise his statutory power to make an award of compound interest if the Contractors have been deprived of their salvage remuneration or special compensation for an excessive period as a result of the Owners' gross misconduct or in other exceptional circumstances.

8.3 If the sum(s) awarded to the Contractors (including the fees and expenses referred to in clause 5.2) are not paid to the Contractors or to the Council by the payment date specified in clause 11.1 the Contractors shall be entitled to additional interest on such outstanding sums from the payment date until the date payment is received by the Contractors or the Council both dates inclusive and at a rate which the Arbitrator shall in his absolute discretion determine in his Award.

9 Currency Correction

In considering what sums of money have been expended by the Contractors in rendering the services and/or in fixing the amount of the Award and/or Appeal Award the Arbitrator or Appeal Arbitrator shall to such an extent and insofar as it may be fair and just in all the circumstances give effect to the consequences of any change or changes in the relevant rates of exchange which may have occurred between the date of termination of the services and the date on which the Award or Appeal Award is made.

10 Appeals and Cross Appeals

10.1 Any party may appeal from an Award by giving written Notice of Appeal to the Council provided such notice is received by the Council no later than 21 days after the date on which the Award was published by the Council.

10.2 On receipt of a Notice of Appeal the Council shall refer the appeal to the hearing and determination of an appeal arbitrator of its choice ("the Appeal Arbitrator").

10.3 Any party who has not already given Notice of Appeal under clause 10.1 may give a Notice of Cross Appeal to the Council within 21 days of that party having been notified that the Council has received Notice of Appeal from another party.

10.4 Notice of Appeal or Cross Appeal shall be given to the Council by letter, e-mail or facsimile.

10.5 If any Notice of Appeal or Notice of Cross Appeal is withdrawn prior to the hearing of the appeal arbitration, that appeal arbitration shall nevertheless proceed for the purpose of determining any matters which remain outstanding.

10.6 The Appeal Arbitrator shall conduct the appeal arbitration in accordance with Lloyd's Procedural Rules so far as applicable to an appeal.

10.7 In addition to the powers conferred on the Arbitrator by English law and the Agreement, the Appeal Arbitrator shall have power to:
   (i)   admit the evidence or information which was before the Arbitrator together with the Arbitrator's Notes and Reasons for his Award, any transcript of evidence and such additional evidence or information as he may think fit;
   (ii)   confirm increase or reduce the sum(s) awarded by the Arbitrator and to make such order as to the payment of interest on such sum(s) as he may think fit;
   (iii) confirm revoke or vary any order and/or declaratory award made by the Arbitrator;
   (iv)   award interest on any fees and expenses charged under clause 10.8 from the expiration of 28 days after the date of publication by the Council of the Appeal Arbitrator's Award until the date payment is received by the Council both dates inclusive.

10.8 The Appeal Arbitrator shall be entitled to satisfactory security for his reasonable fees and expenses, whether such fees and expenses have been incurred already or are reasonably anticipated. The Appeal Arbitrator shall have the power to order one or more of the parties to provide such security in a sum or sums and in a form to be determined by the Appeal Arbitrator. The said power may be exercised from time to time as the Appeal Arbitrator considers appropriate.

10.9 The Appeal Arbitrator's Award shall be published by the Council in London.

11 Provisions as to Payment

11.1 When publishing the Award the Council shall call upon the party or parties concerned to pay all sums due from them which are quantified in the Award (including the fees and expenses referred to in clause 5.2) not later than 28 days after the date of publication of the Award ("the payment date")

11.2 If the sums referred to in clause 11.1 (or any part thereof) are not paid within 56 days after the date of publication of the Award (or such longer period as the Contractors may allow) and provided the Council has not received Notice of Appeal or Notice of Cross Appeal the Council shall realise or enforce the security given to the Council under clause 4.5 by or on behalf of the defaulting party or parties subject to the Contractors' providing the Council with any indemnity the Council may require in respect of the costs the Council may incur in that regard.

11.3 In the event of an appeal and upon publication by the Council of the Appeal Award the Council shall call upon the party or parties concerned to pay the sum(s) awarded. In the event of non-payment and subject to the Contractors providing the Council with any costs indemnity required as referred to in clause 11.2 the Council shall realise or enforce the security given to the Council under clause 4.5 by or on behalf of the defaulting party.

11.4 If any sum(s) shall become payable to the Contractors in respect of salvage remuneration or special compensation (including interest and/or costs) as the result of an agreement made between the Contractors and the Owners or any of them, the Council shall, if called upon to do so and subject to the Contractors providing to the Council any costs indemnity required as referred to in clause 11.2 realise or enforce the security given to the Council under clause 4.5 by or on behalf of that party.

11.5 Where (i) no security has been provided to the Council in accordance with clause 4.5 or (ii) no Award is made by the Arbitrator or the Appeal Arbitrator (as the case may be) because the parties have been able to settle all matters in issue between them by agreement the Contractors shall be responsible for payment of the fees and expenses referred to in clause 5.2. Payment of such fees and expenses shall be made to the Council within 28 days of the Contractors or their representatives receiving the Council's invoice failing which the Council shall be entitled to interest on any sum outstanding at UK Base Rate prevailing on the date of the invoice plus 2% per annum until payment is received by the Council.

11.6 If an Award or Appeal Award directs the Contractors to pay any sum to any other party or parties including the whole or any part of the costs of the arbitration and/or appeal arbitration the Council may deduct from sums received by the Council on behalf of the Contractors the amount(s) so payable by the Contractors unless the Contractors provide the Council with satisfactory security to meet their liability.

11.7 Save as aforesaid every sum received by the Council pursuant to this clause shall be paid by the Council to the Contractors or their representatives whose receipt shall be a good discharge to for it.

11.8 Without prejudice to the provisions of clause 4.5 the liability of the Council shall be limited to the amount of security provided to it.

12 Awards

12.1 The Council will ordinarily make available the Award or Appeal Award and Reasons on www.lloydsagency.com (the website) except where the Arbitrator or Appeal Arbitrator has ordered, in response to representations by any party to the Award or Appeal Award, that there is a good reason for deferring or withholding them. Any party may apply to make such representations to the Arbitrator provided a written notice of its intention to do so is received by the Council no later than 21 days after the date on which the Award or Appeal Award was published by the Council.

12.2 Subject to any order of the Arbitrator or Appeal Arbitrator, the Award, or Appeal Award, and Reasons will be made available on the website as soon as practicable after expiry of the 21 day period referred to in clause 12.1.

12.3 In the event of an appeal being entered against an Award, the Award and Reasons shall not be made available on the website until either the Appeal Arbitrator has issued his Appeal Award or the Notice of Appeal is withdrawn subject always to any order being made in accordance with clause 12.1.

Special Provisions

These Special Provisions shall apply to salved cargo insofar as it consists of laden containers.

13 The parties agree that any correspondence or notices in respect of salved cargo which is not the subject of representation in accordance with Clause 7 of these Rules may be sent to the party or parties who have provided salvage security in respect of that property and that this shall be deemed to constitute proper notification to the owners of such property.

14 Subject to the express approval of the Arbitrator, where an agreement is reached between the Contractors and the owners of salved cargo comprising at least 75% by value of salved cargo represented in accordance with Clause 7 of these Rules, the same agreement shall be binding on the owners of all salved cargo who were not represented at the time of the said approval.

15 Subject to the express approval of the Arbitrator, any salved cargo with a value below an agreed figure may be omitted from the salved fund and excused from liability for salvage where the cost of including such cargo in the process is likely to be disproportionate to its liability for salvage.

General Provisions

16 **Lloyd's documents:** Any Award notice authority order or other document signed by the Chairman of Lloyd's or any person authorised by the Council for the purpose shall be deemed to have been duly made or given by the Council and shall have the same force and effect in all respects as if it had been signed by every member of the Council.

17 **Contractors' personnel and subcontractors**

17.1 The Contractors may claim salvage on behalf of their employees and any other servants or agents who participate in the services and shall upon request provide the Owners with a reasonably satisfactory indemnity against all claims by or liabilities to such employees servants or agents.

17.2 The Contractors may engage the services of subcontractors for the purpose of fulfilling their obligations under clauses A and B of the Agreement but the Contractors shall nevertheless remain liable to the Owners for the due performance of those obligations.

17.3 In the event that subcontractors are engaged as aforesaid the Contractors may claim salvage on behalf of the subcontractors including their employees servants or agents and shall, if called upon so to do provide the Owners with a reasonably satisfactory indemnity against all claims by or liabilities to such subcontractors their employees servants or agents.

18 **Disputes under Scopic Clause**

Any dispute arising out of the Scopic Clause (including as to its incorporation or invocation) or the operations thereunder shall be referred for determination to the Arbitrator appointed under clause 5 hereof whose Award shall be final and binding subject to appeal as provided in clause 10 hereof.

19 **Lloyd's Publications**

Any guidance published by or on behalf of the Council relating to matters such as the Convention the workings and implementation of the Agreement is for information only and forms no part of the Agreement.

Annex 5

LLOYD'S STANDARD FORM OF SALVAGE AGREEMENT

(Approved and Published by the Council of Lloyd's)

PROCEDURAL RULES

(pursuant to Clause I of LOF 2000)

1 Arbitrators Powers

In addition to all powers conferred by the Arbitration Act 1996 (or any amendment thereof) the Arbitrator shall have power:

a to admit such oral or documentary evidence or information as he may think fit;

b to conduct the arbitration in such manner in all respects as he may think fit subject to these Procedural Rules and any amendments thereto as may from time to time be approved by the Council of Lloyd's ("the Council");

c to make such orders as to costs, fees and expenses including those of the Council charged under clauses 5.2 and 10.8 of the Lloyd's Standard Salvage and Arbitration Clauses ("the LSSA clauses") as may be fair and just;

d to direct that the recoverable costs of the arbitration or of any part of the proceedings shall be limited to a specified amount;

e to make any orders required to ensure that the arbitration is conducted in a fair and efficient manner consistent with the aim to minimise delay and expense and to arrange such meetings and determine all applications made by the parties as may be necessary for that purpose;

f to conduct all such meetings by means of a conference telephone call if the parties agree;

g on his own initiative or on the application of a party to correct any award (whether interim provisional or final) or to make an additional award in order to rectify any mistake error or omission provided that (i) any such correction is made within 28 days of the date of publication of the relevant award by the Council (ii) any additional award required is made within 56 days of the said date of publication or, in either case, such longer period as the Arbitrator may in his discretion allow.

2 Preliminary Meeting

a Within 6 weeks of being appointed or so soon thereafter as may be reasonable in the circumstances, the Arbitrator shall convene a preliminary meeting with the represented parties for the purpose of giving directions as to the manner in which the arbitration is to be conducted.

b The Arbitrator may dispense with the requirement for a preliminary meeting if the represented parties agree a consent order for directions which the Arbitrator is willing to approve. For the purpose of obtaining such approval, the Arbitrator must be provided by the contractors or their representatives with a brief summary of the case in the form of a check list, any other party providing such comments as they deem appropriate so that the Arbitrator is placed in a position to decide whether to approve the consent order.

c In determining the manner in which the arbitration is to be conducted the Arbitrator shall have regard to:
  i the interests of unrepresented parties;
  ii whether some form of shortened and/or simplified procedure is appropriate including whether the arbitration may be conducted on documents only with concise written submissions;
  iii the overriding objectives set out in clause 2 of the LSSA clauses.

3 Order for Directions

Unless there are special reasons, the initial order for directions shall include:-

a a date for disclosure of documents including witness statements (see Rule 4);

b a date for proof of values;

c a date by which any party must identify any issue(s) in the case which are likely to necessitate the service of pleadings;

d a date for a progress meeting or additional progress meetings unless all represented parties with reasonable notice agree that the same is unnecessary;

e unless agreed by all represented parties to be premature, a date for the hearing and estimates for the time likely to be required by the Arbitrator to read evidence in advance and for the length of the hearing;

f any other matters deemed by the Arbitrator or any party to be appropriate to be included in the initial order.

4 Disclosure of documents

Unless otherwise agreed or ordered, disclosure shall be limited to the following classes of document:

a logs and any other contemporaneous records maintained by the shipowners personnel and personnel employed by the Contractors (including any subcontractors) and their respective surveyors or consultants in attendance during all or part of the salvage services;

b working charts, photographs, video or film records;

c contemporaneous reports including telexes, facsimile messages or prints of e- mail messages;

d survey reports;

e documents relevant to the proof of:
  i out of pocket expenses
  ii salved values
  iii the particulars and values of all relevant salving tugs or other craft and equipment

f statements of witnesses of fact or other privileged documents on which the party wishes to rely.

5 Expert Evidence

a No expert evidence shall be adduced in the arbitration without the Arbitrators permission.

b The Arbitrator shall not give such permission unless satisfied that expert evidence is reasonably necessary for the proper determination of an issue arising in the arbitration.

c No party shall be given permission to adduce evidence from more than one expert in each field requiring expert evidence save in exceptional circumstances.

d Any application for permission to adduce expert evidence must be made at the latest within 14 days after disclosure of relevant documents has been effected.

6 Mediation

The Arbitrator shall ensure that in all cases the represented parties are informed of the benefit which might be derived from the use of mediation.

7 Hearing of Arbitration

a In fixing or agreeing to a date for the hearing of an arbitration, the Arbitrator shall not unless agreed by all represented parties fix or accept a date unless the Arbitrator can allow time to read the principal evidence in advance, hear the arbitration and produce the award to the Council for publication in not more than 1 month from conclusion of the hearing.

b The date fixed for the hearing shall be maintained unless application to alter the date is made to the Arbitrator within 14 days of the completion of discovery or unless the Arbitrator in the exercise of his discretion determines at a later time that an adjournment is necessary or desirable in the interests of justice or fairness.

c Unless all parties represented in the arbitration agree otherwise the Arbitrator shall relinquish his appointment if a hearing date cannot be agreed, fixed or maintained in accordance with rule 7(a) and/or (b) above due to the Arbitrator's commitments. In that event the Council shall appoint in his stead another arbitrator who is able to meet the requirements of those rules.

8 Appeals

a All references in these Rules to the Arbitrator shall include the Arbitrator on Appeal where the circumstances so permit.

b In any case in which a party giving notice of appeal intends to contend that the Arbitrator's findings on the salved value of all or any of the salved property were erroneous, or that the Arbitrator has erred in any finding as to the person whose property was at risk, a statement of such grounds of appeal shall be given in or accompanying the notice of appeal.

c In all cases grounds of appeal or cross-appeal will be given to the Arbitrator on Appeal within 21 days of the notice of appeal or cross- appeal unless an extension of time is agreed.

d Any respondent to an appeal who intends to contend that the award of the Original Arbitrator should be affirmed on grounds other than those relied upon by the Original Arbitrator shall give notice to that effect specifying the grounds of his contention within 14 days of receipt of the grounds of appeal mentioned in (c) above unless an extension of time is agreed.

Annex 6

SUB-CONTRACT (AWARD SHARING) 2001

BETWEEN

The Contractor

– and –

The Sub-Contractor

THIS AGREEMENT is made the   day of   2

Between:-

for and on behalf of ("the Contractor")

and

for and on behalf of ("the Sub-Contractor").

WHEREAS:

(1)   The Contractor is presently or is about to become engaged in rendering salvage services to the "   ", her cargo, freight, bunkers, stores and any other property thereon under a Lloyd's Standard Form of Salvage Agreement "No Cure - No Pay" dated   ("the LOF") and the Contractor wishes to engage the services of the Sub-Contractor on "No Cure - No Pay" terms to assist him in the performance of his obligations under the LOF.

(2)   The Sub-Contractor is willing to assist the Contractor and will provide the personnel, equipment and services set out in the schedule, together with any further personnel, equipment or services which may reasonably be requested by the Contractor from time to time during the performance of the services.

NOW in consideration of the mutual promises and undertakings contained herein it is agreed as follows:-

DEFINITIONS

1.   In this Agreement the following expressions shall have the following meanings:-

(a)   "ISU Terms" means award sharing terms which are the same (mutatis mutandis) or substantially the same as the terms of this agreement;

(b)   "Non Award Sharing Terms" means terms which provide for the remuneration of a sub-contractor otherwise than by way of a share of the Salvage Remuneration payable under the LOF;

(c)   "Relevant Sub-Contract" means a sub-contract whereby the Contractor engages the service of a sub-contractor on ISU Terms to assist him in the performance of his obligations under the LOF;

(d)   "Salvage Remuneration" means any and all remuneration plus interest thereon paid or payable whether awarded, agreed or received in respect of services rendered under the LOF, including out of pocket expenses, interim payments, article 14 special compensation, SCOPIC Remuneration and costs.

OBLIGATIONS OF CONTRACTOR

2.   The Contractor agrees:-

(a)   To engage the Sub-Contractor on "No Cure - No Pay" terms to assist him in the performance of his obligations under the LOF and to include in his claim for Salvage Remuneration the agreed services rendered by the Sub-Contractor;

(b)   To share with the Sub-Contractor the Salvage Remuneration as finally awarded under the LOF or as agreed by the parties thereto and received by the Contractor;

(c)   To use his best endeavours to recover, as part of those legal costs as finally awarded to him under the LOF, or as agreed between the parties thereto, such legal costs (calculated on the standard basis) incurred by the Sub-Contractor in and about the provision and presentation of evidence by him for use in the LOF proceedings, and thereafter to account to the Sub-Contractor for his legal costs recovered as aforesaid;

(d)   As soon as is reasonably practical to instruct solicitors in London to open an identified interest bearing client deposit account established in accordance with the Solicitors Act 1974 (as amended) (the "Trust Account") and to hold in the Trust Account any and all such remuneration as may be received pursuant to the LOF as stakeholder on trust for the Contractor and the Sub-Contractor with authority, upon final agreement or determination of the share due to each party, to release such share on being requested in writing by such party so to do; and

(e)   As soon as reasonably practicable to inform the Council of Lloyd's and the owners of the salved property and the guarantors of the existence of this Agreement, the identity of the Sub-Contractor and to give them irrevocable instructions to pay any and all monies due under the LOF to the Trust Account.

OBLIGATIONS OF SUB-CONTRACTOR

3.   The Sub-Contractor agrees:-

(a)   To use his best endeavours to assist the Contractor in the performance of his obligations under the LOF, including the provision of such personnel, equipment and services as are set out in the attached schedule or as are reasonably requested by the Contractor during the performance of the service;

(b)   To assist the Contractor in the presentation of the claim for Salvage Remuneration by the provision and retention of evidence relating to the salvage services and to the sub-contractor's contribution to the same;

(c)   Not to claim Salvage Remuneration and/or article 14 special compensation against the owner(s) of the property salved or any part thereof, nor to make any claim for Salvage Remuneration in respect of the services rendered pursuant to this agreement, save insofar as this agreement provides; and

(d)   To provide an indemnity satisfactory to the Contractor against any successful claim for Salvage Remuneration made by the sub-contractor's servants and/or agents and/or sub-contractors (and/or the servants and/or agents of the same) against the owner(s) of such property.

TRUST

4.

(a)   From the time when this agreement is concluded, all sums paid or payable by way of Salvage Remuneration due under the LOF or this agreement shall be owned in law by the Contractor and the Sub-Contractor jointly and shall be subject to a trust in favour of the Contractor and Sub-Contractor as beneficiaries. Save as is expressly provided in this agreement, neither party shall have the right to assign or otherwise dispose of or deal with such sums or any part Thereof or any interest therein.

(b)   In the event of either party receiving any sum howsoever on account of or in payment of Salvage Remuneration in respect of any of the services rendered under the LOF or this agreement, such sum shall be held on trust as aforesaid for the Contractor and the Sub-Contractor and shall forthwith be paid into the Trust Account.

CONDUCT OF LOF ARBITRATION

5   (a)   The Arbitration under the LOF and/or any negotiations for an amicable settlement shall be conducted solely by the Contractor but the Sub-Contractor agrees to provide all necessary evidence and assistance in connection therewith. The Contractor shall keep the Sub-Contractor fully advised as to the amount of the security demanded from the Owner(s) of the salved property and as to the nature and form of the guarantees received and the identity of all guarantors.

(b)   The Contractor, so far as the circumstances reasonably permit, shall consult with the Sub-Contractor and keep the Sub-Contractor informed at all significant stages of the Arbitration or of any settlement negotiations, but failing agreement by the Sub-Contractor the Contractor shall be entitled at his discretion to proceed with the Arbitration or to conclude a bona fide settlement.

(c)   The Sub-Contractor shall be entitled to attend the Arbitration as an observer but at his own cost.

ASSIGNMENT

6   (a)   The Contractor shall not without the consent in writing of the Sub-Contractor (such consent not to be unreasonably withheld) make or purport to make any assignment of the benefit of the LOF or of the whole or any part of the Salvage Remuneration.

(b)   Neither the Contractor nor the Sub-Contractor shall without the consent in writing of the other (such consent not to be unreasonably withheld) make or purport to make any assignment of the benefit of this agreement or of the share of the Salvage Remuneration to which it is entitled under the joint operation of the LOF and this agreement or of its interest under the trust herein before contained.

INDEMNITIES

7   (a)   The Contractor agrees to indemnify fully and hold harmless the Sub-Contractor, his servants and/or agents against any claim by the owner(s) of the property salved or by any other person (other than the Sub-Contractor's own servants and/or agents) for loss or damage caused by the negligence in the salvage operations of the Contractor, his servants or agents and/or by any defects (other than latent defects) in any equipment of the Contractor used in the salvage operation.

(b)   The Sub-Contractor agrees to indemnify fully and hold harmless the Contractor, his servants and/or agents against any claim by the owner(s) of the property salved or by any other person (other than the Contractor's own servants and/or agents) for loss or damage caused by the negligence in the salvage operation of the Sub-Contractor, his servants or agents and/or by any defects (other than latent defects) in any equipment of the Sub-Contractor used in the salvage operation.

(c)   The Contractor will make no claim against the Sub-Contractor, his servants and/or agents for loss or damage sustained by the Contractor's equipment or by any servant or agent of the Contractor caused by the negligence of the Sub-Contractor, his servants or agents and/or by defects in the Sub-Contractor's equipment and hereby agrees to indemnify fully and hold harmless the Sub-Contractor in respect of any such claim made by the Contractor's servants and/or agents.

(d) The Sub-Contractor will make no claim against the Contractor, his servants and/or agents for loss or damage sustained by the Sub-Contractor's equipment or by any servant or agent of the Sub-Contractor caused by the negligence of the Contractor, his servants and/or agents and/or by defects in the Contractor's equipment and hereby agrees to indemnify fully and hold harmless the Contractor in respect of any such claim made by the Sub-Contractor's servants and/or agents.

LIMITATION OF LIABILITY

8. Notwithstanding anything contained herein, and in particular clause 7, either party to this agreement shall be entitled to limit any liability to the other party which he and/or his servants and/or agents may incur in and about the services under this agreement in the manner and to the extent provided by English law, save that this clause shall not apply to any liability under clause 3(d) above.

TRUSTEE EXONERATION

9 Neither the Contractor nor any other trustee of the trust herein before contained shall be liable for or for the consequences of any error or mistake (whether by way of commission or omission and whether on the part of the Contractor or other trustee himself, or on the part of any agent or adviser employed or instructed by the Contractor or other trustee) made or committed in or about the agreement or ascertainment or recovery of the Salvage Remuneration, or the obtaining enforcement or release of any security therefor or otherwise in or about the execution of the trust herein before contained unless such error or mistake shall be proved to have occurred or been committed in personal conscious bad faith of the party sought to be made liable.

RELATIONSHIP TO OTHER SUB-CONTRACTS

10 a) Save as may be specifically advised, the Contractor hereby warrants that he has not hitherto engaged the services of any other party to assist him in the performance of his obligations under the LOF except upon ISU Terms or Non-Award Sharing Terms.

(b) The Contractor may at any time hereafter engage the services of such other parties as he may think fit to assist him in the performance of his obligations under the LOF, but the terms upon which each such party is engaged must either be ISU Terms or Non-Award Sharing Terms.

(c) Where in relation to the performance of his obligations under the LOF, the Contractor has entered or hereafter enters into any engagement with another party on ISU Terms or Non Award Sharing Terms he will, at the request of the Sub-Contractor, furnish to the Sub-Contractor all information in his possession concerning the identity and address of every such other party.

(d) In case this agreement is not the only Relevant Sub-Contract, the following provisions shall apply:-

(i) Clause 2(d) (release of money from the Trust Account pursuant to written direction), Clause 4 (declaration of trust by this agreement) and clause 11 (quantification of the shares of the Contractor and the Sub-Contractor in the Salvage Remuneration) shall have effect as if references to every Sub-Contractor under all Relevant Sub-Contracts were substituted for references to the Sub-Contractor;

(ii)   The Contractor will so instruct solicitors in London that a single account becomes the Trust Account in relation to every relevant Sub-Contract;

(iii)   Whenever a matter or dispute has to be determined or resolved by arbitration pursuant to this agreement and an arbitrator has already been appointed to determine or resolve the same, or an equivalent matter or dispute under another Relevant Sub-Contract, then notwithstanding anything to the contrary in clause 11(a) of this agreement, that matter or dispute shall be referred to the same arbitrator (if he so consents) and so far as possible he will determine or resolve that matter in relation to every Relevant Sub-Contract in a single arbitration.

ARBITRATION

11   (a)   Any dispute arising hereunder, including any dispute as to the share of the Salvage Remuneration due to the Contractor and Sub-Contractor, shall be referred to the arbitrament of an arbitrator, to be selected by the first party claiming arbitration from the persons currently on the panel of Lloyd's salvage arbitrators with a right of appeal from an award made by the arbitrator to either party by notice in writing to the other within 28 days of the date of publication of the original arbitrator's award.

(b)   The arbitrator on appeal shall be the person currently acting as Lloyd's appeal arbitrator.

(c)   Both the arbitrator and the appeal arbitrator shall have the same powers as an arbitrator and appeal arbitrator respectively would have under LOF 2000 or any standard revision thereof, including a power to order a payment on account of a share due to a party pending final determination of any dispute between the parties hereto.

(d)   For the purposes of any apportionment between the parties the term "Salvage Remuneration" as defined in clause 1(d) shall include all expenses, unrecovered costs and brokers' commissions. The arbitrator or arbitrator on appeal, unless he/she considers it unfair or unjust so to do, in determining the shares of the aforesaid Salvage Remuneration, shall first apportion to the Contractor and/or Sub-Contractor(s) all sums specifically awarded other than the sums for Salvage Remuneration and interest under LOF and all reasonably incurred out of pocket expenses, unrecovered but customary costs and brokers' commissions. Where interest has been awarded or earned in respect to any of the foregoing by reason of the LOF award or award on appeal, and/or by reason of being held in the Trust Account, the said interest shall be included in the initial apportionment.

RELEVANT LAW

12.   This Agreement shall be governed and construed in accordance with English law.

AS WITNESS the hands of the duly authorised representatives of the parties hereto.

For and on behalf of the Contractor   For and on behalf of the Sub-Contractor

....

SCHEDULE OF PERSONNEL, EQUIPMENT AND SERVICES

1. Date and Place of Agreement:	DAILY HIRE AGREEMENT
	S A L V H I R E 2 0 0 5 **PART I**
2. Hirer; Place of Business:	3. Owner; Place of Business: (Part II - Clause 1.3)

4. Detail and Specification of Vessel hired under this Agreement; (Part II - Preamble and Clauses 1.2, 1.3, 2 and 26).

5. Name of Owner's P&I Association:

6. Details of Casualty; (Part II - Preamble, Clauses 1.1, 6.1 and 6.3)

a) Name:

b) Flag:

c) Place of Registry:

d) Owners:

e) Length:

f) Beam:

g) Maximum draft:

h) Displacement:

i) Details and Nature of Cargo:

j) Any other Casualty's details relevant to this Agreement:

7. Condition of Casualty:

8. Location of Casualty:

9. Nature of Services to be provided by the Owner: (Part II - Clauses 6.1, 10 and 13.3)

10. Vessel Rates of Hire; (including Bunkers, Lubricating Oil and Water): (Part II - Clauses 12.1 and 13.1)

a) Working Rate, (not towing):
b) Towing Rate:
c) Standby Rate: (i) At Anchor -   (ii) At Sea -   (iii) In Port -

11. Mobilisation/Demobilisation Fee, (if applicable): (Part II - Clauses 12.1 and 13.1)

12. Hire to commence from: (Part II - Clause 2 and 13.2)	13. Hire to terminate at: (Part II - Clause 2 and 13.2)
14. Minimum Number of Days Hire: (Part II - Clauses 2, 12.1 and 13.1)	15. Maximum duration of hire: (Part II - Clauses 2, 3 and 13.2)

16. Payment Details: (Part II - Cláuse 13.5)

Currency:
Bank:
Address:
Sort Code:
Account Number:
Account Name:
Reference:

17. Time for Payment and Interest: (Part II - Clause 14)

Monies not paid within   calendar days of presentation of the Owner's invoice shall attract interest of   percent per month.

18. Extra Costs: (Part II - Clause 15.2)	19. Security Requirements: (Part II - Clauses 13.6, 16.1 and 16.2)
Handling Charge of percent to be applied.	

20. Law and Arbitration: (Part II - Clauses 24 and 25)

Arbitration to take place at :
If this Box left blank then Part II, Clause 24.1 shall apply.

21. Number of Additional Clauses:

The undersigned warrant that they have full power and authority to sign this Agreement on behalf of the parties represented by them. In the event of a conflict of terms and conditions, the provisions of Part I and any additional clauses, if agreed, shall prevail over those of Part II to the extent of such conflict but no further.

....

FOR AND ON BEHALF OF THE HIRER

....

FOR AND ON BEHALF OF THE OWNER

CODE OF PRACTICE BETWEEN INTERNATIONAL SALVAGE UNION
AND INTERNATIONAL GROUP OF P&I CLUBS

In the spirit of co-operation, the following Code of Practice is agreed between the International Salvage Union and the International Group of P&I Clubs in relation to all future salvage services to which Article 14 of the 1989 Salvage Convention is applicable or under Lloyd's Form where the Special Compensation P&I Club's (SCOPIC) Clause has been invoked by the Contractor.

1. The salvor will advise the relevant P&I Club at the commencement of the salvage services, or as soon thereafter as is practicable, if they consider that there is a possibility of a Special Compensation claim arising.

2. In the event of the SCR not being appointed under the SCOPIC clause, the P&I Club may appoint an observer to attend the salvage and the salvors agree to keep him and/or the P&I Club fully informed of the salvage activities and their plans. However, any decision on the conduct of the salvage services remains with the salvor.

3. The P&I Club, when reasonably requested by the salvor, will immediately advise the salvor whether the particular Member is covered, subject to the Rules of the P&I Club, for any liability which he may have for Special Compensation or SCOPIC Remuneration.

4. The P&I Clubs confirm that, whilst they expect to provide security in the form of a Club Letter either in respect of claims for special compensation (under Article 14 of the 1989 Salvage Convention) or SCOPIC remuneration (under the SCOPIC Clause), as appropriate, it is not automatic. Specific reasons for refusal to give security to the Contractor will be non-payment of calls, breach of warranty rules relating to classification and flag state requirements or any other breach of the rules allowing the Club to deny cover. The Clubs will not refuse to give security solely because the Contractors cannot obtain security in any other way.

5. In the event that security is required by a port authority or other competent authority for potential P&I liabilities in order to permit the ship to enter a port of refuge or other place of safety, the P&I Clubs confirm that they would be willing to consider the provision of such security subject to the aforementioned provisos referred to in para. 4 above and subject to the reasonableness of the demand.

6. The Contractors will accept security for either special compensation or SCOPIC remuneration by way of a P&I Club letter of undertaking in the attached form - "Salvage Guarantee form – ISU 5" - and they will not insist on the provision of security at Lloyd's.

7. The P&I Club concerned will reply to any request by the salvors regarding security as quickly as reasonably possible. In the event that salvage services are being performed under Lloyd's Form incorporating the SCOPIC clause, the P&I Club concerned will advise the Contractor within two (2) working days of his invoking the SCOPIC Clause whether or not they will provide security to the Contractor by way of a Club Letter referred to in para. 6 above.

8. In the event that salvage services are being performed under Lloyd's Form incorporating the SCOPIC clause, the P&I Clubs will advise the owners of the vessel not to exercise the right to terminate the contract under SCOPIC Clause 9(ii) without reasonable cause.

9. It is recognised that any liability to pay SCOPIC remuneration is a potential liability of the shipowner and covered by his liability insurers subject to the Club Rules and terms of entry. Accordingly, in the event of such payment of SCOPIC remuneration in excess of the Article 13 award, neither the shipowner not his liability insurers will seek to make a claim in General Average against the other interests to the common maritime adventure whether in their own name or otherwise and whether directly or by way of recourse or indemnity or in any other manner whatsoever.

10. The P&I Clubs, if consulted, and the ISU will recommend to their respective Members the incorporation of the SCOPIC clause in any LOF.

11. The P&I Clubs and the ISU will not agree to any variation of the terms, including the tariff rates, of the SCOPIC Clause except in accordance with the provisions of the SCOPIC Clause itself, particularly Appendix B.

12. This is a Code of Practice which the ISU and the International Group of P&I Clubs will recommend to their Members and it is not intended that it should have any legal effect.

1.1999
).2000
.2005
.6.2005

Summary

Firsthand experience of a marine emergency cannot be gained from the text on the pages of a book. What to do when at the sharp end can only be learned from that real-time situation of being there and living through the experience. It is hoped that the reader will never be in such an incident as to have to learn firsthand. However, so many of us, as seafarers, find ourselves in that one-off scenario, when you may need to make a judgement, hopefully the right decision at the right time.

The purpose of this text is not to replace that live experience, as valuable as it is, but to provide food for thought, in the event of influencing a correct decision when needed. Knowing that in an emergency situation a wrong decision in our profession can lead to death or serious injury for crew members and passengers alike.

The collision at sea, the grounding, that serious fire on board, are all real enough, but few mariners get to see them all at once. The text has tried to show the influencing factors which lead to the emergency. They will never ever be the same. The weather will be different, the ship involved will be a different vessel and the geographical position will be the ultimate variable when it goes wrong for the individual.

Society can only educate to accommodate for that time, that day, when the individual makes that decision which he believes to be the right one. The chapters of this text, with the case studies and recorded incidents, have hopefully covered a broad spectrum of a senior marine officer's emergency duty.

The work throughout has been based on that first principle of 'safety of life at sea' being paramount, through any and every voyage.

Good Sailing and fair weather to all,
D.J. House, 2013

Bibliography

SOLAS Consolidated Edition (2009), IMO.

House, D.J. (2002), *Anchor Practice: A Guide to Industry*, Witherby.

House, D.J. (2005), *Marine Heavy Lifting & Rigging Operations*, Brown Son & Ferguson Ltd.

House, D.J. (2011), *Marine Survival* (3rd edn), Witherby.

House, D.J. (2014), *Seamanship Techniques* (4th edn), Routledge.

IMO (2013), *IAMSAR Manual*, volume III, IMO.

Maclachlan, M. (2013), *The Shipmaster's Business Self Examiner*, Nautical Institute.

The Mariners Handbook (9th edn) (2009), Admiralty.

MARPOL (International Convention for the Prevention of Pollution from Ships) (2012), Amendments to Annexes I, II, IV, V and VI.

Index

Abandonment 7, 19, 105–136, 201–204
Abandonment miscellaneous facts
    134–135
Abandonment psychology 113–114
Accommodation fire 95
AMVER 192
Anchor:
    Cable / chain 67, 72
    Deep water 64
    Dragging 64–67
    Handling vessel xix
    Kedge 67, 70–71
    Plan 63
    Reaction 67–70
    Stern 62, 67, 70–71
    Warp xix, 67, 71
Anchor use 59–72
Arbitration xix
Arbitrator xix
Archaeological salvage xix
Auto pilot 54

Backstays xix
Ballast 4, 58
Ballast water movement / management
    159–160
Barge transports 169–171
Beaching xix, 7, 25, 32, 199
Beaching conditions 32–33
Beaufort wind scale 75
Bibliography 237
Bitter end xx
Boats 115–126
Bollard pull xx, 164–166
Boundary cooling 7, 85, 88–89, 97
Braer, M.V. (loss) 78–79
Breathing apparatus 85, 89, 177, 205
Bridge interface 54
Broken stowage xx
Bulk / Bulk carriers xx, 92
Bulk chemical code 138
Bull wire xx
Bunga Alpinia (fire) 104
Bunker checklist 157

Cable shackle xxvi, 72
Cable strength 72
Calypso (fire) 118
Carbon Dioxide CO_2 system / flooding
    86–88

Cargo barges 169–171
Cargo hold fires 91–94
Cargo salvage xx
Cargo shift xx, 28, 76
Carpenters stopper xx
Cast Beaver (grounding) 35
Casualties 6, 13
Cement Box xx
Certificates:
    Class 15, 22
    Of approval xxi
    Of fitness 138
    Passenger Ship Safety 17
    Safety equipment 86
    Seaworthiness 15
Checklist use 7–8, 192–193
Chemical tanker 138–139
Clean ballast 139
Closed loop control 53
Coal Cargo 91, 210
Cofferdam xxi, 139
Collision:
    Casualties 6
    Chief Officers role 12
    Communications 5–6, 9
    Flooding 4
    Immediate actions 1–2
    Information exchange 5
    Legal actions 5
    Masters role 5–6
    Obligations 5
    Passenger ship 17–19
    Patch / mat xxi, 13
    Position 6
    Scenarios 3– 4
    Tanker 19–20
Combination carrier 139
Compass 51–52, 55
Compass reliability 51–52
Composite towline xxi, 168
Co-ordinated search pattern 196
Costa Concordia (loss) 107–113
Cradle xxi
Crude oil 139

Damage assessment 3, 5, 12, 27, 34,
    199–200
Damage control parties 3, 7, 16, 85
Damage report 20–22
Davit launched liferafts 123–124

Deadweight xxi
Deck security 75
Delayed turn 185
Derelict xxi
Designated person ashore (DPA) 157, 195–196
Displacement xxii
Distress procedures 192–193, 195
Distress signals 196–198
Docking types 40–46
Double Elliptical Turn 185
Drills 2, 81, 127–128
Dry docking 25, 39–47, 98, 99
Dynamometer xxii

ECDIS 187–188
Electro-Hydraulic steering 51
Emergency:
    Communications 194–196, 200
    Docking 39–40
    Drills 127–128
    Equipment locker 17
    Escape Breathing Devices 89
    Generator 57
    SAR plan (passengers) 8
    Steering xxii
    Towing arrangement xxii, 161, 165, 167, 169
Empress of Canada (fire) 81
Enclosed spaces 86, 89, 92, 177–178
Engine, loss of 57
Engine room fires 87–90
EPIRB / GPIRB 31, 198
Estonia (capsize) 17–18
Evacuation by:
    Davit launched raft 123–124
    Free fall boat 120
    Helicopter 128–129
    Inflatable liferaft 121–122
    Marine Evacuation system 126–127
Even Keel xxii
Exposure 118–119
Exxon Valdez (pollution) 152–153

Fire: 6, 81–104, 204–206
    Accommodation 94–97
    Alarm 82
    At sea / in port 81–82, 83
    Container vessel 101–102
    Conventional fighting 88
    Examples 87–104
    Foam monitor 103
    Galley 96
    In dock 98–99
    Inert gas use 96–97
    Master's actions 82
    Monitor xxii

    On deck 101
    Paint locker 97
    Safety design 104
    Station 100
    Support units 84, 175
    Tanker collision / fire 19, 102–104
    Teams 85, 103
    Triangle 87
First aid party 84
Flag effect 58
Flaminia (explosion / fire) 92
Floating dock xxii, 41–42
Flooding 6, 17, 112
Flotsam xxii
Flow Moisture Point xxii
Fog encounter 178–181
Free fall lifeboats 120
Free surface xxii–xxiii

Galley fire 96
Garbage 140
Gas carrier 140
Gas free certificate 99
General alarm 12
General average xxiii, 12, 15–16, 91
Girding (Girting) xxiii
Girdling xxiii
GMDSS 3, 200
Gog xxiii
Grounding xxiii, 25–39, 187, 199–200
Ground tackle xxiii, 67, 71, 154
Guest warp xxiii, 192

Harmful substances 140
Heave to xxiii, 77–78
Heavy weather 73
Heavy weather ship handling 77
Helicopter activity 128–135, 211
Helicopter landing officer 131
Herald of Free Enterprise (capsize) xxvi, 106–107
Hi-Line operation 133–134
Holding ground xxiii
Hull stresses, grounding / docking 47
Hydrolift docking 45–46
Hydrostatic release unit 122, 135
Hypothermia 119

IAMSAR 188–189, 207–208
IBC code 140
Ice navigation 181–183
Ice Patrol 199
IGC code 140
Ignition Point 140
IMDG code 82
IMO 142
Immersion risk 118–119

Impact damage 4
Inert gas use 96–97
Insurance (marine) xxiv
Insurance risk 169
Interim Certificate of Class 15, 22, 39
International Safety Management xxiv, 7, 195
International Salvage Union, sub-contract 227
ISGOTT 137–138
ISU contract xx

Jetsam xxiv
Jury sea anchor 57
Jury steering xxii, xxiv, 57

Kedge anchor xxiv, 67, 70–71
Kilindo rope xxiv
Kulluk drilling rig (grounding) 158

Lagan xxiv
Lead line 37–38
Lee shore 50, 57
Legal actions in collision 5
Legal representation 201
Lifeboats 115–116, 120–121
Liferafts 121–124, 208
Lifting beam xxiv
Lightening operations 153–154
List 4
Lloyds Agency xxiv
Lloyds Agreement, Clauses 219
Lloyds Agreement, procedural rules 225
Lloyds Open Form (of salvage) xxiv–xxv, 177, 213–227
Load density plan xxv
Loss of control 57
Lost buoyancy xxv
Luffing xxv
Lutine Bell xxv

Maanan Star (grounding) 27
Man Overboard 183–187, 188–190
Marine Evacuation system 126–127
Marine incidents 211
Marine pilot transfers 186–187
Marine Pollution Control Unit 141
Marine Rescue Co-ordination Centre 191–192, 193
Marine safety information 194–195
Mayday 6–7, 9
Media contact 202
Medivac 132–133
Miscellaneous hazards 177–198
Mousing xxv

Napoli (beaching) 31–32

NAVTEX 194–195
New dangers 207
Not Under Command xxv, 58
Noxious Liquid Substance Certificate 141

Oil:
    Movement 157
    Pipelines 156–157
    Record book 138
    Recovery 158–159
    Skimmer xxv
    Spill 78, 151–152
    Tanker 142
On Scene Co-ordinator 197
Open loop control 52
Overhauling xxv

Paint locker fire 97
Parbuckle xxv, 111
Partially enclosed lifeboat 116
Particular Average xxiii, xxvi
Passage Plan 79, 187
Passenger behaviour 18–19, 113–118
Passenger muster 117
Passenger ship fires 99–100
Permeability xxvi
Permit to work xxvi, 86
Pilot boat 186
Pilot transfers 186–187
Pollutants 145
Pollution 137–160
Pollution boom xxvi, 33, 35, 138
Pollution causes 146
Port of refuge 7,14, 92
Port of refuge, communications 15
Pounding 77
Power loss 6, 57
Product Carrier 142
Proof Load xxvi
Protest, Note of 11–12, 91
Pumps 4, 112–113

Rams Horn hook xxvi
Receiver of Wreck xxvi
Release Note xxvi
Repair assessment 20–22
Rescue boat 124, 126, 184, 186
Risk assessment xxvi
Riverdance (grounding) xxi, 26, 28
Rope Gauge xxvi
Ro-Ro cargo spaces 93
Roystan Grange (collision / fire) 19

Safe Working Load xxvi
Safety contour 188
Safety Depth 187
Sally Albatross 26, 105

Salvage: xxi, xxvii
  Agreement xxvii
  Association xxvii
  Contact 173–176
  Contractual xxi
  Convention xxiv
  Craft xxvii
  Cutting operations xxi
  Information quality 176
  Lien xxvii
  Life xxvi
  Operations 68–69
  Underwater xxix
Salvor xxvii
SAR communications 198
SAR operations 190, 192, 193, 206
SAR Plan 190, 197
SART 31, 198, 202
Search Patterns 184, 188–190
Seaworthiness xxvii
Sector Search189–190
Security 195–196
Segregated ballast 143
Sewage 143
Shackle of Cable xxvii, 72
Sheer Legs xxvii, 172–173
Shell Expansion Plan 22
Shipboard Marine Pollution Emergency
      Plan (SEMEP) 143
Ship to ship transfer xxviii
Ship types xxx
Ships Oil Pollution Emergency Plan
      (SOPEP) 137, 143, 150
Skeg xxviii
Slop tank 143
Sounding xxviii, 37, 180, 197
Special Area 143–144
Sprinkler system 93, 97–98
Steering:
  Control 52–54
  Electro / Hydraulic 51
  Failure 55
  Four Ram 51
  Jury xxiv
  Loss of 50, 55, 57–58
  Rotary Vane 56
  Transmission 54
Stowage factor xxviii
Stranded xxiii, 12, 16, 25
Sulphur 92
Summary 235

Survival equipment 192
Synchro-Lift docking 43–44
Synchronised motion 73

Tanker 144, 147–150, 164
Tanker terminology 138–145
Territorial waters xxviii
Tien Chee (collision) 19
Towing:
  Emergency 165
  Line xxviii, 162, 165–167
  Operations 161–168
  Option 7
  Point xxviii
  Spring xxviii
Tow master 164
Toxic leak 101, 204
Track space 189, 190–191
Training 82, 187, 203
Tricolour xxi
Trimming xxix
Tugs:
  Approval survey xxix, 166
  Ocean going xxv
  Operations 162–170

Uhmpe xxviii
Ullage xxix, 145
Ulster Sportsman (grounding) 26
Underway xxix
Underwriter xxix
Urgency signal 6,7, 9, 58, 82, 195, 197
US Coast Guard 193–194

Ventilation party 84
Visibility poor 178–180
Visibility table 180
Void space 145
Voyage Data Recorder 9

Warranty survey xxix
Warranty surveyor xxix
Watchkeeping 179–180
Water:
  Curtains 93
  Ingress 4
  Mist systems 90, 205
  Tight doors 206
Williamson Turn 184
Windlass capability 72
Wreck xxix

Milton Keynes UK
Ingram Content Group UK Ltd.
UKHW051926141024
449569UK00027B/1375